CELL BIOCHEMISTRY
A QUICK REVIEW

CELL BIOCHEMISTRY
A QUICK REVIEW

Edited by

Dr. Vijaykumar Deshmukh

Delve
PUBLISHING

www.delvepublishing.com

Cell Biochemistry: A Quick Review

Dr. Vijaykumar Deshmukh

Delve Publishing
4164 Lakeshore Road
Burlington, ON L7L 1A4
Canada
www.delvepublishing.com
Email: orders@arclereducation.com

ISBN: 978-1-77956-362-0 (Hardcover)

ABOUT THE EDITOR

Dr. Vijaykumar D Deshmukh is presently serving as Assistant Professor in Department of Plant Biotechnology, MGM College of Agricultural Biotechnology, Chh. Sambhajinagar (Maharashtra). He obtained his M.Sc. (Agri. Biotech.) in 2005 and Ph.D. (Biotech.) in 2022 from Dr. D.Y. Patil University, Pune (Maharashtra). He started his career as an Assistant Professor in 2005. His field of specialisation is Plant Biotechnology and Molecular Biology. He has published five books and more than 25 research papers in National and international journals. He has participated in International, National and State Level Seminar and Symposiums.

TABLE OF CONTENTS

LIST OF FIGURES

ABSTRACT

"Cell Biochemistry" explores the intricate biochemical processes that underpin cellular functions. This comprehensive text delves into the molecular foundations of cell structure, metabolism, and signaling pathways. It covers essential topics such as enzyme kinetics, membrane dynamics, and energy production, emphasizing the interplay between biochemical reactions and cellular physiology. By integrating theoretical concepts with experimental techniques, the book equips readers with a deep understanding of how biochemical processes govern life at the cellular level. Ideal for students and professionals alike, it serves as a crucial resource for those seeking to comprehend the biochemical mechanisms that drive cellular behavior and functionality.

PREFACE

This book on cell biochemistry delves into the molecular mechanisms that sustain cellular function, focusing on the four main macromolecules: proteins, nucleic acids (deoxyribonucleic acid [DNA] and ribonucleic acid [RNA]), lipids, and carbohydrates. Each chapter explores the structural and functional properties of these biomolecules, revealing their roles in cellular processes. Proteins, the most versatile macromolecules, are highlighted for their enzymatic and structural roles, which drive a multitude of biochemical reactions required for cellular function. Nucleic acids, such as DNA and RNA, are fundamental molecules that store, transmit, and express genetic information, orchestrate protein synthesis, and regulate gene function. Lipids are studied for their roles in forming cellular membranes, creating barriers that regulate the cell's internal and external environments, and participating in signaling pathways. Carbohydrates are described not only as primary energy sources but also as key players in cellular recognition and signaling mechanisms.

The book provides an in-depth description of cellular metabolism, detailing the metabolic pathways that convert nutrients into energy, essential for cellular maintenance and function. It covers glycolysis, the citric acid cycle, and oxidative phosphorylation, with an emphasis on how these pathways are regulated by enzyme activity, feedback mechanisms, and cellular signals. The study of cell biochemistry is further enriched with advanced topics such as structural biology, systems biology, and molecular genetics, all of which contribute to a complete understanding of cellular processes.

This textbook was inspired by the desire to bridge fundamental biochemical knowledge with recent scientific advances. It seeks to provide a comprehensive understanding of cellular function, emphasizing both traditional biochemical pathways and new discoveries that have transformed the field. The combination of structural biology and high-throughput omics technologies, such as genomics and proteomics, reflects the evolution of cell biochemistry, in which traditional concepts are continuously reshaped by new methodologies and insights. By focusing on both established principles and emerging trends, this book aims to provide students and researchers with the knowledge and tools they need to navigate and contribute to the dynamic field of cell biochemistry.

In compiling this book, the author was motivated by the recognition that an in-depth understanding of cell biochemistry is essential not only for biologists and biochemists but also for those in medical and health sciences. The molecular basis of cellular function has far-reaching implications for understanding disease mechanisms, developing therapeutic strategies, and advancing biotechnology. The author hopes that by delving into the molecular details of cellular life, readers will gain a greater appreciation for the complexity and beauty of biochemical systems, fostering a generation of scientists prepared to face future biological challenges.

This textbook serves as a thorough introduction for newcomers to the field and a valuable reference for experienced researchers. It is designed to be accessible while maintaining scientific rigor, allowing readers from various backgrounds to develop a solid understanding of cell biochemistry. The book offers a detailed exploration of the chemical foundations of life, preparing readers to engage with and contribute to the rapidly evolving field of molecular biology.

Chapter 1

Introduction to Cell Biochemistry

Contents

Within molecular biology, cell biochemistry is a basic field that studies the chemical mechanisms governing cellular structure and function. This field studies the complex interactions between molecules that control cellular life, with a particular emphasis on the four main classes of biomolecules: proteins, lipids, carbohydrates, and nucleic acids. These macromolecules are what make up cells, and they are all vital for preserving cellular integrity and promoting the biochemical processes that are necessary for life (Baynes and Dominiczak, 2009) (Figure 1.1).

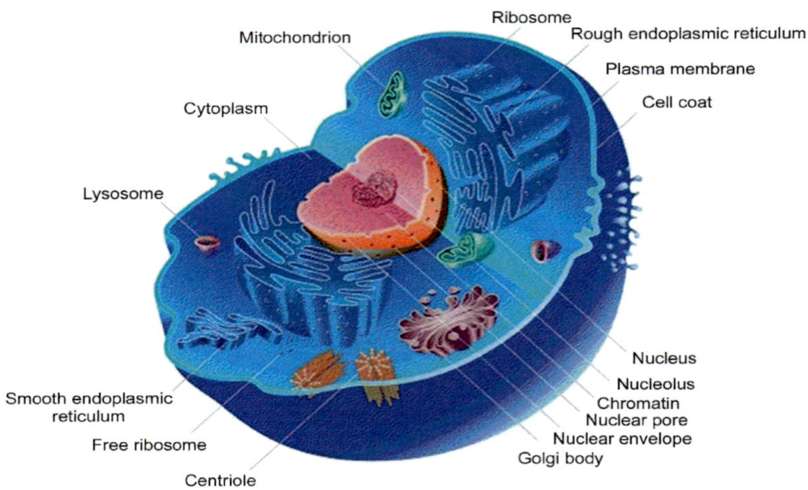

Figure 1.1. *The animal cell.*

Source: https://courses.lumenlearning.com/hccs-waymakerbiology1/chapter/animal-cell-structure/.

Proteins, which are made up of amino acids, are arguably the most versatile macromolecules found in cells. They function as enzymes, catalyzing biochemical reactions with high specificity and efficiency, as well as structural components that contribute to the cytoskeleton and extracellular matrix. Protein structure determines function, a principle central to understanding cellular processes at the molecular level (Boal, 2012). The specificity of enzyme-substrate interactions, as well as the regulation of enzymatic activity, are important topics in cell biochemistry, as they underpin the diverse metabolic pathways that sustain cellular life.

Nucleic acids, such as DNA and RNA, serve as both storage and transmission sites for genetic information. Watson and Crick's discovery of

DNA's double helical structure provided the blueprint for all cellular functions. The replication, transcription, and translation processes, which ensure the accurate transmission and expression of genetic information, are fundamental concepts in cell biochemistry. RNA molecules, including messenger RNA (mRNA), transfer RNA (tRNA), and ribosomal RNA (rRNA), play critical roles in protein synthesis and gene regulation (Boal, 2012; Bonner and Varner, 2012).

Lipids, particularly phospholipids, are critical components of cellular membranes, forming bilayers that compartmentalize the cell and create distinct microenvironments required for various biochemical reactions. Membrane dynamics, including the fluid mosaic model and the role of lipid rafts in signaling, are critical for understanding how cells maintain homeostasis and respond to external stimuli (Bonner and Varner, 2012).

Carbohydrates play a role in cell recognition and signaling in addition to providing energy. Glycoproteins and glycolipids on cell surfaces play critical roles in cell-cell communication and immune response. Carbohydrates' structural diversity, ranging from simple monosaccharides to complex polysaccharides, is critical to their function in biological systems (Boal, 2012).

Another important aspect of cell biochemistry is the study of cellular metabolism, which includes the complex networks of metabolic pathways that convert nutrients into energy and biomolecules. Bioenergetics, the study of energy flow within cells, relies heavily on pathways like glycolysis, the citric acid cycle, and oxidative phosphorylation. These pathways are tightly regulated, with feedback mechanisms ensuring that cellular energy requirements are met efficiently (Boal, 2012). Cell biochemistry advances, particularly through the integration of structural biology, systems biology, and molecular genetics, are helping us better understand cellular processes. X-ray crystallography, nuclear magnetic resonance (NMR) spectroscopy, and cryo-electron microscopy have all provided previously unknown insights into biomolecule structure-function relationships (Baynes and Dominiczak, 2009). Furthermore, the development of high-throughput omics technologies such as genomics, proteomics, and metabolomics has enabled comprehensive analysis of cellular systems, revealing life's complexity and dynamism at the molecular level.

1.1. THE CELL: BASIC UNIT OF LIFE

The cell is universally recognized as the fundamental unit of life. At its core, the cell is a complex network of interconnected biochemical processes that provide life. This complexity is reflected in the cell's diverse array of macromolecules, all of which contribute to its structure, function, and dynamic nature (Alberts et al., 2015).

The plasma membrane is the cell's distinguishing feature, a lipid bilayer made up primarily of phospholipids, cholesterol, and proteins. This membrane not only provides structural integrity but also regulates the selective permeability of substances, allowing homeostasis to be maintained. Various proteins are embedded in this membrane, allowing the cell to communicate with its environment, transport molecules, and mediate signal transduction pathways. The fluid mosaic model of the membrane emphasizes its dynamic nature, allowing lipids and proteins to move laterally and adapt to changing conditions (Alberts et al., 2015).

Inside the cell, the cytoplasm contains a complex network of organelles, each with a unique biochemical function. The nucleus, surrounded by a double membrane, houses the cell's genetic material in the form of DNA. This genetic blueprint is required for the regulation of cellular activities such as replication, transcription, and translation, which are critical to gene expression and protein synthesis (Boal, 2012). Proteins are the cell's workhorses, serving as enzymes, structural components, and signaling molecules.

The endoplasmic reticulum (ER) and Golgi apparatus play critical roles in protein synthesis, folding, and trafficking. The ER, particularly the rough ER with its ribosome-studded surface, is responsible for protein synthesis, whereas the smooth ER is involved in lipid synthesis and detoxification processes. The Golgi apparatus further processes and sorts these proteins and lipids before transporting them to their final destinations (Baynes and Dominiczak, 2009).

Mitochondria, also known as the cell's powerhouses, are the sites of oxidative phosphorylation, which produces ATP, the cell's primary energy currency. This energy is required for a variety of cellular processes, including muscle contraction and active transport across membranes (Baynes and Dominiczak, 2009). Mitochondria also play a role in apoptosis, the programmed cell death needed to maintain tissue homeostasis.

The cytoskeleton, composed of microfilaments, microtubules, and intermediate filaments, provides structural support and facilitates intracellular transport, cell division, and motility. These dynamic structures are involved in processes such as endocytosis, where cells engulf external substances and cytokinesis, the division of the cytoplasm during cell division (Alberts et al., 2015).

Cellular communication is another essential aspect of cell biology. Chemical signals, such as hormones and neurotransmitters, are used by cells to communicate. These signals bind to specific receptors on the cell surface, activating signal transduction pathways that cause a cellular response. This communication is critical for coordinating activities within multicellular organisms, ensuring that cells function in a harmonious and regulated manner (Boal, 2012).

1.2. OVERVIEW OF CELL STRUCTURE

The cell, as the basic unit of life, has a highly organized and complex structure that allows it to perform a variety of biochemical functions required for survival, growth, and replication. The structural organization of the cell is inextricably linked to its function, with each component playing a unique and critical role.

The plasma membrane is a dynamic, selectively permeable barrier that defines the cell's boundaries. It consists primarily of a lipid bilayer made up of phospholipids, cholesterol, and glycolipids, with embedded proteins that contribute to its functional diversity (Csordás et al., 2006). The amphipathic nature of phospholipids, with hydrophobic tails and hydrophilic heads, promotes the formation of the bilayer, resulting in a stable barrier that separates the intracellular and extracellular spaces.

The fluid mosaic model of the plasma membrane describes how lipids and proteins move laterally within the bilayer, allowing for flexibility and the formation of specialized microdomains like lipid rafts. These rafts are high in cholesterol and sphingolipids, and they act as platforms for signal transduction and protein sorting (Boal, 2012). Integral membrane proteins, such as channels, receptors, and transporters, help the cell communicate with its surroundings, regulate the passage of ions and molecules, and mediate signal transduction pathways.

The cytoplasm, which includes the cytosol and organelles, is another component of the cell, and a critical site for numerous biochemical reactions that keep cells alive. The cytosol, or aqueous component of the cytoplasm, is a complex solution containing water, ions, small molecules, and macromolecules like proteins, nucleotides, and carbohydrates. It serves as the medium for metabolic pathways such as glycolysis and the pentose phosphate pathway (Bonner and Varner, 2012).

The cytoplasm also acts as a buffer, maintaining the pH and ionic composition necessary for enzymatic activities. Additionally, it plays a role in intracellular signaling, with second messengers like cyclic AMP and calcium ions being distributed throughout the cytoplasm to propagate signals (Alberts et al., 2015).

The cell also contains organelles, which are membrane-bound structures that perform specific biochemical functions contributing to the cell's overall physiology. Organelle compartmentalization allows for the separation of incompatible biochemical processes and the creation of specialized environments for specific reactions, The nucleus is the cell's control center, containing genetic material in the form of chromatin, a complex of DNA and histone proteins. The nuclear envelope, a double membrane structure that surrounds the nucleus, is perforated by nuclear pores that control material exchange between the nucleus and the cytoplasm (Alberts et al., 2015). The nucleolus is a substructure of the nucleus that synthesizes ribosomal RNA (rRNA) and assembles ribosomes.

The nucleus is where critical processes like DNA replication, transcription, and RNA processing occur. Transcription factors and other regulatory proteins in the nucleus regulate gene expression, which influences cellular function and identity (Bonner and Varner, 2012).

The ER is a network of membranous tubules and sacs responsible for protein and lipid synthesis, folding, modification, and transport. The rough ER, which is studded with ribosomes, is the site of protein synthesis for proteins that will be secreted into the plasma membrane or lysosomes. The smooth ER, which lacks ribosomes, performs lipid synthesis, detoxification, and calcium storage (Baynes and Dominiczak, 2009).

Molecular chaperones and folding enzymes tightly regulate protein folding within the ER, ensuring that only properly folded proteins reach the Golgi apparatus. The ER-associated degradation (ERAD) pathway targets misfolded proteins for degradation, preventing the accumulation of dysfunctional proteins (Csordás et al., 2006).

The Golgi apparatus, which is made up of stacked, membrane-bound cisternae, is in charge of modifying, sorting, and packaging proteins and lipids obtained from the ER. Glycosylation, phosphorylation, and sulfation are three post-translational modifications that occur within the Golgi and are required for protein function and trafficking (Alberts et al., 2015). The Golgi is also important for the formation of lysosomes and the secretion of cellular products.

Mitochondria are double-membrane organelles that serve as the cell's powerhouse, producing ATP via oxidative phosphorylation. The inner mitochondrial membrane is highly folded into cristae, increasing the surface area of the electron transport chain (ETC) complexes and ATP synthase, which are responsible for ATP production. Mitochondria are also involved in apoptosis, calcium homeostasis, and the synthesis of specific lipids and amino acids (Boal, 2012).

Mitochondrial DNA (mtDNA), inherited maternally, encodes several ETC components, emphasizing the organelle's distinct semi-autonomous nature (Baynes and Dominiczak, 2009). The relationship between the nuclear and mitochondrial genomes is critical for mitochondrial function and cellular energy metabolism.

Lysosomes are acidic organelles containing hydrolytic enzymes that degrade macromolecules, damaged organelles, and cellular debris via autophagy and endocytosis. They are essential for maintaining cellular homeostasis and defending against pathogens (Alberts et al., 2015).

Peroxisomes are small, single-membrane organelles that catabolize fatty acids, detoxify reactive oxygen species, and synthesize plasmalogens, which are essential components of myelin. The enzyme catalase in peroxisomes converts hydrogen peroxide, a byproduct of fatty acid oxidation, into water, thereby preventing oxidative damage (Bonner and Varner, 2012).

The cytoskeleton is a dynamic network of protein filaments that provides structural support, facilitates intracellular transport, and enables cellular movement. It is composed of three main types of filaments: microfilaments, intermediate filaments, and microtubules.

Microfilaments, composed of actin, are the thinnest cytoskeletal filaments and are involved in cell motility, shape, and division. Actin filaments interact with myosin to generate contractile forces, playing a critical role in muscle contraction, cytokinesis, and the maintenance of cell shape (Baynes and Dominiczak, 2009).

Intermediate filaments, made up of proteins like keratins, vimentin, and Lamins, give cells mechanical strength and structural stability. They form a network that spans the cytoplasm and nucleus, supporting the nuclear envelope and ensuring the integrity of tissues under mechanical stress (Alberts et al., 2015).

Microtubules are hollow cylinders composed of α and β tubulin dimers. They are essential for intracellular transport, mitosis, and maintaining cell shape. Motor proteins such as kinesin and dynein use microtubules as tracks for organelles and vesicles to move along. During mitosis, chromosomes are separated into daughter cells by the mitotic spindle, which is made up of microtubules (Csordás et al., 2006; Bonner and Varner, 2012).

1.3. FUNCTIONS OF CELLULAR ORGANELLES

The nucleus is the distinguishing organelle of eukaryotic cells, often referred to as the control center due to its role in storing the cell's genetic material. It is surrounded by the nuclear envelope, which is a double membrane that separates the nucleus from the cytoplasm and controls the passage of molecules through nuclear pores. DNA is organized into chromatin within the nucleus, where it condenses to form chromosomes when cells divide (Csordás et al., 2006). The nucleus' primary functions are to store and protect genetic information, regulate gene expression, and coordinate cell division. The nucleus controls the transcription of DNA into messenger RNA (mRNA), which is then translated into proteins in the cytoplasm. This process is tightly regulated by a complex network of transcription factors and epigenetic modifications, which ensure that genes are expressed in a controlled and timely manner, essential for the cell's response to internal and external stimuli (Figure 1.2).

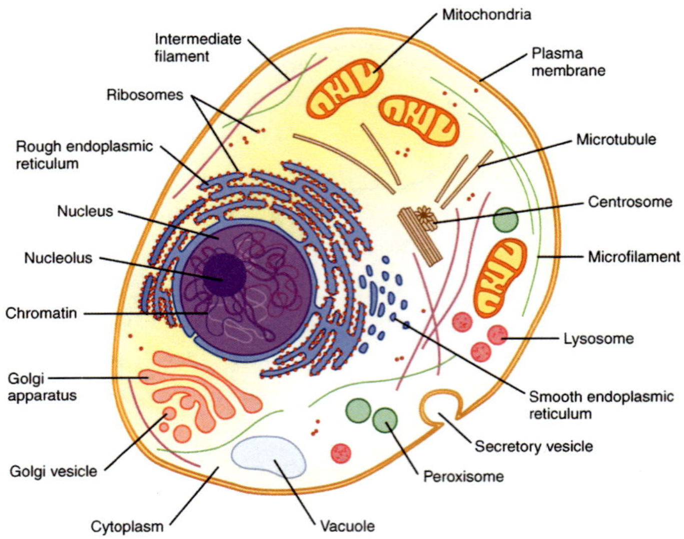

Figure 1.2. *Cell organelles.*

Source: https://courses.lumenlearning.com/suny-ap1/chapter/the-cytoplasm-and-cellular-organelles/.

The endoplasmic reticulum (ER) is a vast network of membranous tubules and sacs responsible for protein and lipid synthesis. The ER is divided into two regions: rough and smooth, each with its own set of functions. The rough ER, which has ribosomes on its cytoplasmic surface, is the site of protein synthesis for secretory proteins, membrane proteins, and lysosomal enzymes. Newly synthesized polypeptides enter the ER lumen, where they are folded, modified (via glycosylation), and quality controlled before being transported to the Golgi apparatus for further processing. The smooth ER, which lacks ribosomes, is involved in the synthesis of lipids, including phospholipids and cholesterol, which are important components of cellular membranes (Alberts et al., 2015).

Additionally, the smooth ER is responsible for detoxifying harmful substances, metabolizing carbohydrates, and storing calcium ions, which are crucial for various cellular processes, including muscle contraction and signal transduction (Baynes and Dominiczak, 2009). The Golgi apparatus is composed of flattened, membrane-bound sacs called cisternae. It serves as the main hub for modifying, sorting, and packaging proteins and lipids synthesized in the ER. Proteins and lipids undergo a variety of post-translational modifications as they pass through the Golgi, including glycosylation, phosphorylation, and sulfation, all of which are necessary for their function and destination. The Golgi apparatus also helps to form lysosomes and secrete proteins via vesicles. By sorting and directing these molecules to their proper cellular locations, the Golgi ensures that proteins and lipids are delivered to the appropriate cellular compartments, allowing proper cellular function (Phillips and Voeltz, 2016).

Mitochondria are double-membraned organelles that primarily produce energy via oxidative phosphorylation. The inner mitochondrial membrane is highly folded into cristae, increasing the surface area for the electron transport chain (ETC) and ATP synthase, the enzymes that produce ATP, the cell's primary energy currency. Mitochondria also play important roles in other cellular processes, such as apoptosis (programmed cell death), metabolism regulation, and the biosynthesis of key molecules like amino acids and lipids (Blumwald et al., 2000). The presence of mitochondrial DNA (mtDNA) demonstrates the organelle's semi-autonomous nature, as it encodes a subset of proteins involved in the ETC, reflecting mitochondria' evolutionary origins as ancestral prokaryotic cells.

Lysosomes are membrane-bound organelles containing hydrolytic enzymes that can degrade a wide range of biomolecules, including proteins, nucleic acids, lipids, and carbohydrates. These enzymes perform best at acidic pH levels, which are maintained within lysosomes, allowing them to efficiently degrade cellular debris, damaged organelles, and macromolecules. Lysosomes are essential for autophagy, which is the process by which cells recycle their components in order to maintain cellular homeostasis and respond to nutrient deprivation (Boal, 2012). Lysosomes also play a role in the immune response, degrading pathogens engulfed by phagocytosis. Lysosomal dysfunction can cause an accumulation of undigested material, resulting in lysosomal storage diseases, which are frequently severe and debilitating.

Peroxisomes are small, membrane-bound organelles that perform oxidation reactions, resulting in the production of hydrogen peroxide, a potentially harmful byproduct that is then broken down by the enzyme catalase. Peroxisomes are involved in the catabolism of very long-chain fatty acids, the detoxification of reactive oxygen species, and the biosynthesis of plasmalogens, all of which are necessary for the nervous system to function properly. Peroxisome enzymes play a crucial role in metabolic pathways, such as the β-oxidation of fatty acids, which is essential for energy production, particularly in cells that rely heavily on fatty acid metabolism. Defects in peroxisomal function can result in peroxisomal biogenesis disorders, such as Zellweger syndrome, which causes severe developmental abnormalities and is frequently fatal (Csordás et al., 2006). The cytoskeleton is a dynamic network of protein filaments that provides structural support, promotes intracellular transport, and allows for cell motility. It consists of three types of filaments: microfilaments, intermediate filaments, and microtubules. Actin-based microfilaments play a role in cell motility, cytokinesis, and shape maintenance. They work with myosin motor proteins to produce contractile forces that are required for muscle contraction and cellular movement. Intermediate filaments help cells maintain mechanical strength and structural stability. They are composed of various proteins, including keratins, vimentin, and lamins, which support the nuclear envelope and maintain tissue integrity, particularly in cells under mechanical stress (Boal, 2012).

Microtubules are hollow cylinders made of α- and β-tubulin dimers that serve as tracks for the movement of organelles, vesicles, and chromosomes during cell division. They play a critical role in maintaining cell shape, intracellular transport, and the formation of the mitotic spindle during mitosis, ensuring the accurate segregation of chromosomes to daughter cells (Phillips and Voeltz, 2016).

1.4. CELL MEMBRANE AND TRANSPORT MECHANISMS

The cell membrane, a dynamic and selectively permeable barrier, is critical in controlling the exchange of materials between the intracellular and extracellular environments. This regulation is critical for maintaining cellular homeostasis, allowing the cell to absorb nutrients, eliminate waste products, and respond to external signals. The cell membrane's transport mechanisms are broadly classified as passive transport, active transport, and bulk transport, with each governed by specific biochemical principles and driven by the cell's metabolic and energetic needs (Phillips and Voeltz, 2016) (Figure 1.3).

Passive transport is the movement of molecules across the cell membrane without the use of cellular energy, owing to the concentration gradient of the molecules involved. This process involves simple diffusion, facilitated diffusion, and osmosis. Simple diffusion occurs when small, nonpolar molecules such as oxygen, carbon dioxide, and lipophilic substances pass directly through the lipid bilayer, moving from higher to lower concentrations. The principle of entropy governs this movement, in which molecules spontaneously distribute to achieve equilibrium (Stein, 2012).

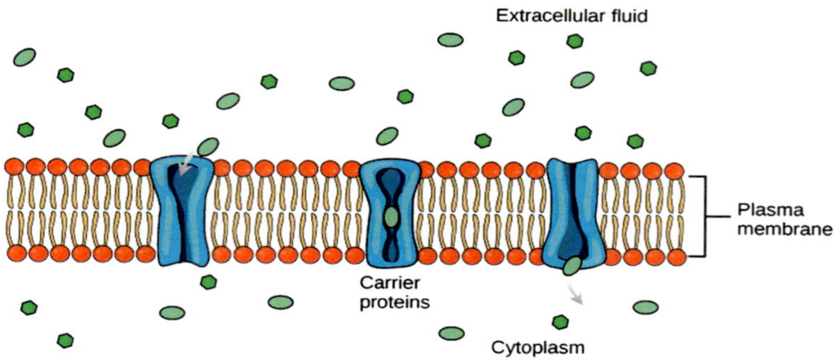

Figure 1.3. *Carrier proteins aid in passive transport across the plasma membrane.*

Source: https://commons.wikimedia.org/wiki/File:0306_Facilitated_Diffusion_Carrier_Protein.jpg.

In contrast, facilitated diffusion refers to the transport of polar molecules and ions that cannot easily pass through the lipid bilayer's hydrophobic core. Specific transmembrane proteins, such as channel proteins and carrier proteins, mediate this process by providing a pathway for these molecules. Channel proteins form hydrophilic pores that allow ions and small molecules to pass through the membrane, whereas carrier proteins change conformation to transport specific molecules across it. The glucose transporter (GLUT) is an example of facilitated diffusion; it allows glucose to enter cells in response to concentration gradients and plays an important role in cellular metabolism (Boal, 2012).

Osmosis is a specific type of passive transport in which water molecules move across a semipermeable membrane. Water moves from areas of low solute concentration (hypotonic) to areas of high solute concentration (hypertonic) to balance solute concentrations on both sides of the membrane. This process is essential for regulating cell volume and osmotic balance. Aquaporins, a family of channel proteins, aid in the rapid and selective transport of water molecules, making them essential in tissues where water permeability is critical, such as the kidneys and red blood cells (Blumwald et al., 2000).

Active transport, unlike passive transport, necessitates the direct expenditure of energy, typically in the form of ATP, to move molecules against a concentration gradient. This process is critical for maintaining concentration gradients of ions and other molecules across the cell membrane, which are required for a variety of cellular functions. Pumps, such as the sodium-potassium pump (Na+/K+ ATPase), play a primary role in active transport by actively transporting sodium ions out of the cell and potassium ions into the cell, thereby maintaining the electrochemical gradients required for nerve impulse transmission and muscle contraction (Stein, 2012). The Na+/K+ ATPase pump is an electrogenic pump that contributes to membrane potential by creating a net charge difference across the membrane.

Secondary active transport, also known as coupled transport or cotransport, is the use of a primary active transport electrochemical gradient to propel the movement of other molecules. In this process, one molecule moves down its gradient while another moves against it. There are two main types of coupled transporters: symporters and antiporters. Symporters, such as the sodium-glucose cotransporter, transport both molecules in the same direction, allowing glucose to enter cells along with sodium ions. Antiporters, such as the sodium-calcium exchanger, transport molecules in opposite directions, and play an important role in regulating intracellular calcium levels, which are critical for various signaling pathways (Blumwald et al., 2000; Stein, 2012).

Bulk transport mechanisms, such as endocytosis and exocytosis, enable the cell to transport large molecules like proteins and polysaccharides, as well as particles, across the membrane. Endocytosis is the process by which the cell membrane engulfs extracellular materials, forming vesicles that are then internalized into the cytoplasm. This process can occur via a variety of pathways, including phagocytosis, in which specialized cells such as macrophages engulf large particles such as pathogens, and pinocytosis, in which the cell consumes extracellular fluid and dissolved solutes. Receptor-mediated endocytosis is a more specific type of endocytosis in which receptors on the cell surface bind to ligands, causing these receptor-ligand complexes to be internalized, such as cholesterol uptake via low-density lipoprotein (LDL) receptors (Bonner and Varner, 2012).

Exocytosis is the reverse process in which intracellular vesicles fuse with the plasma membrane and release their contents into the extracellular space. This process is essential for hormone, neurotransmitter, and enzyme secretion, as well as membrane component recycling. Regulated exocytosis of neurotransmitters at synaptic terminals is critical for nerve signal transmission, whereas constitutive exocytosis is responsible for the continuous secretion of molecules like collagen in the extracellular matrix (Alberts et al., 2015).

1.5. CELLULAR METABOLISM

Cellular metabolism, a fundamental element of cell biology, refers to the metabolic processes by which cells transform nutrients into energy and building blocks for growth and survival. This complex network of metabolic pathways not only maintains cellular function but also plays an important role in health and illness. It is typically classified into two types: catabolism and anabolism. Catabolic routes degrade complex molecules into simpler ones, producing energy that is commonly collected in the form of adenosine triphosphate (ATP). In contrast, anabolic processes employ energy to produce complex molecules from simpler substrates. The balance of catabolism and anabolism is critical for cellular homeostasis and function (Boal, 2012).

Glycolysis, the citric acid cycle (Krebs cycle), and oxidative phosphorylation are the three main metabolic mechanisms. A tiny quantity of ATP is produced during the cytoplasmic process of glycolysis, which reduces glucose to pyruvate while also producing reducing equivalents in the form of NADH (Nelson et al., 2008). After entering the mitochondria, pyruvate undergoes further oxidation in the citric acid cycle. More NADH and FADH2 are produced by this cycle and are fed into the inner mitochondrial membrane's electron transport chain (ETC). The major mechanism for ATP synthesis by oxidative phosphorylation is driven by the ETC (Alberts et al., 2015).

1.5.1. Regulation of Metabolism and Its Implications in Disease

Metabolic pathways within cells are systematically controlled to respond to changing energy needs and nutritional availability. This regulation is critical for maintaining cellular homeostasis and involves a variety of methods, including enzyme activity modulation, allosteric regulation, and covalent modifications. For example, glycolysis and gluconeogenesis, which are essential for glucose metabolism, are regulated by important enzymes like hexokinase and phosphofructokinase. In glycolysis, for example, phosphofructokinase-1 (PFK-1) is vulnerable to feedback inhibition by high levels of ATP, thereby delaying glycolysis when energy supplies are adequate (Phillips and Voeltz, 2016). When ATP levels are low, PFK-1 activity is raised to boost glycolysis, allowing the cell to satisfy its increased energy needs.

Similar to this, the availability of substrate and the need for reducing equivalents determine how the citric acid cycle is regulated. Acetyl-CoA, NAD+, and other metabolic intermediate quantities affect important enzymes including citrate synthase and isocitrate dehydrogenase. This control makes sure the cycle runs smoothly in a range of cellular environments. Moreover, the availability of oxygen and the presence of particular inhibitors or uncouplers that affect the efficiency of ATP synthesis influence the electron transport chain (ETC), which produces ATP (Csordás et al., 2006). It takes these regulatory mechanisms to balance energy output with the demands of the cell. Disruptions in these metabolic pathways are often linked to various diseases, highlighting the importance of metabolic regulation in health. Cancer, for instance, is frequently associated with altered metabolic processes, a phenomenon known as the Warburg effect. Cancer cells exhibit a preference for aerobic glycolysis over oxidative phosphorylation, even in the presence of adequate oxygen. This metabolic shift supports rapid cell proliferation and provides intermediates required for biosynthetic processes essential for tumor growth (Warburg, 1956). Targeting these metabolic alterations has emerged as a promising therapeutic strategy in oncology, aiming to exploit the unique metabolic vulnerabilities of cancer cells (Alberts et al., 2015).

Glycolysis and gluconeogenesis are two metabolic pathways that are impacted by elevated blood glucose levels in the context of diabetes, especially type 2 diabetes, which is caused by impaired glucose metabolism and insulin resistance. According to Boal, (2012), this dysregulation is a contributing factor to serious long-term consequences such as neuropathy and cardiovascular disease. To manage and reduce these consequences, therapeutic approaches—which include pharmacological therapies and lifestyle modifications—focus on enhancing insulin sensitivity and controlling glucose metabolism.

Neurodegenerative diseases, such as Alzheimer's disease, also reflect the consequences of metabolic dysfunction. Mitochondrial impairments and disrupted glucose metabolism in the brain are prominent features of these conditions, suggesting that metabolic interventions may play a role in managing these diseases (Phillips and Voeltz, 2016). Ongoing research into the metabolic pathways involved in neurodegeneration continues to explore potential therapeutic targets, aiming to address the metabolic aspects of these debilitating disorders.

Overall, metabolic control is a complex and critical element of cellular function, with far-reaching consequences for disease processes and treatment approaches. Understanding how metabolic pathways are controlled and how dysregulation leads to various diseases allows researchers to develop tailored therapies to better manage and treat these conditions. This ongoing study emphasizes the importance of metabolic control in promoting health and treating disease.

1.5.2. Emerging Research and Technologies

Recent breakthroughs in metabolomics, or the thorough investigation of metabolites inside biological systems, are giving new information about cellular metabolism and its role in disease. Mass spectrometry and nuclear magnetic resonance (NMR) spectroscopy have improved the capacity to accurately identify and measure metabolites. Metabolomic research has revealed biomarkers for a variety of disorders and enhanced our understanding of metabolic changes linked to pathological situations (Hewitt, 2004). Furthermore, research into cellular metabolism is increasingly emphasizing the importance of mitochondrial function and dynamics. Mitochondria are responsible for energy production, apoptosis, cellular signaling, and the creation of reactive oxygen species (ROS). Mitochondrial malfunction has been linked to a variety of disorders, and new medicines that target mitochondrial metabolism are being developed (Phillips and Voeltz, 2016).

1.6. IMPORTANCE OF CELL BIOCHEMISTRY IN MEDICINE

Cell biochemistry is a fundamental discipline in the biological sciences, giving vital insights into the molecular mechanisms that govern cellular function, disease pathophysiology, and therapeutic interventions. Its impact on medicine is enormous, affecting everything from disease diagnosis and treatment to the creation of personalized treatments and improved diagnostic tools. Understanding disease mechanisms on a cellular level is critical for developing successful treatments. Many diseases, including cancer, neurological disorders, and metabolic syndromes, result from aberrant cellular processes. For example, cancer is frequently characterized by uncontrolled cell proliferation due to abnormalities in genes that regulate the cell cycle and apoptosis. The exploration of these biochemical pathways has led to the identification of key molecular targets for cancer therapies, including tyrosine kinases and checkpoint inhibitors, significantly improving patient outcomes through targeted treatments (Hewitt, 2004).

In drug development, cell biochemistry is critical for finding and verifying pharmacological targets. The interactions between medications and biological biomolecules—such as enzymes, receptors, and transporters—are critical in biochemical study. By understanding these molecular interactions, high-throughput screening techniques and structural biology have accelerated the discovery of new medication candidates (Hewitt, 2004). For example, the creation of Gleevec (imatinib), a targeted therapy for chronic myeloid leukemia, was influenced by a thorough biochemical understanding of the BCR-ABL fusion protein and its function in disease pathophysiology. This example demonstrates the relevance of cell biochemistry in developing particular and effective medicinal medicines.

Personalized medicine makes extensive use of cell biochemistry to customize treatments to individual genetic and biochemical profiles. Understanding how genetic variants alter biochemical processes and medication reactions is useful in developing personalized methods. Biomarkers that predict patient responses to various medications have been identified through genomic investigations and biochemical analyses. For example, the discovery of HER2 overexpression in breast cancer prompted the adoption of trastuzumab (Herceptin) as a targeted therapy, dramatically improving patient outcomes (Jain et al., 2021). This shift toward individualized therapy demonstrates how cell biology improves clinical decision-making.

The creation and improvement of diagnostic instruments are fundamentally based on biochemical concepts. Biochemical assays are used in many diagnostic procedures to quantify particular molecules or biochemical events. Based on biochemical principles, methods like polymerase chain reaction (PCR) and enzyme-linked immunosorbent assays (ELISAs) are frequently employed for pathogen and biomarker detection (Murray et al., 2016). Diagnostic tests for disorders including cardiovascular diseases (high-sensitivity C-reactive protein test) and prostate cancer (PSA test) have been developed as a result of biochemical research's identification of biomarkers. These instruments are essential for tracking the advancement of the disease, determining the effectiveness of treatment, and detecting it early.

Gene therapy represents a cutting-edge application of cell biochemistry, focusing on correcting or replacing defective genes responsible for diseases. This approach relies on a detailed understanding of gene expression, regulation, and delivery mechanisms. Techniques such as CRISPR/Cas9 gene editing, based on biochemical principles, have shown promise in treating genetic disorders by precisely modifying DNA sequences. The development of gene therapy strategies necessitates in-depth knowledge of cellular processes to ensure accurate and safe delivery of therapeutic genes, expanding treatment possibilities for previously incurable genetic diseases (Murray et al., 2016).

Cell biochemistry is crucial to the research of metabolic pathways and their dysregulation in metabolic disorders in the fields of metabolism and nutritional medicine. Studies on the citric acid cycle and glycolysis have shed light on the role that metabolic abnormalities have in the development of diseases such as lipid disorders, diabetes, and obesity (Jain et al., 2021). As a result of this understanding, nutritional treatments and metabolic medicines have been developed to address or regulate these imbalances, demonstrating the usefulness of cell biochemistry in improving public health.

The development of vaccines heavily relies on cell biochemistry. Comprehending the pathogen-host cell dynamic and the immune system's reaction is crucial for the development of vaccines. Through the identification of important antigens and the optimization of their presentation to the immune system, biochemical investigations of bacterial and viral proteins have made it easier to produce successful vaccines. As per Murray et al. (2016), the progress made in biochemical and molecular biology research has fueled the creation of mRNA vaccines against COVID-19, including those produced by Pfizer-BioNTech and Moderna. These vaccinations serve as prime examples of how cell biochemistry can be used to address issues related to global health.

As a result, cell biochemistry is a fundamental component of modern medicine, giving critical insights into disease causes, medication discovery, tailored therapeutics, diagnostics, gene therapy, metabolism, and vaccine design. The incorporation of biochemical research into medical practice has resulted in important advances in understanding and treating numerous diseases. As our understanding of cell biochemistry grows, its applications in medicine will definitely expand, opening up new avenues for improving human health and addressing complicated medical challenges (Hewitt, 2004).

1.7. FUTURE TRENDS IN CELL BIOCHEMISTRY RESEARCH

Driven by technological advancements and a deeper understanding of cellular mechanisms, future trends in cell biochemistry research are poised to change our understanding of cellular processes and their consequences for health and disease. Owing to advances in imaging tools, cell biochemistry is being transformed by never-before-seen insights into cellular dynamics and architecture. Super-resolution microscopy is one technique that allows researchers to see cellular processes at the nanoscale scale. It goes beyond the diffraction limit of light. The precision of studying protein interactions and cellular architecture has already been improved by stimulated emission depletion (STED) microscopy and single-molecule localization microscopy (SMLM). Future developments are anticipated to further improve spatial and temporal resolution, enabling real-time visualization of biochemical processes at the single-molecule level. These innovations promise a more detailed understanding of cellular dynamics and the molecular mechanisms underlying disease (De Mandal et al., 2022).

Single-cell omics technologies are also transforming the understanding of cellular heterogeneity and function. Traditional biochemical assays frequently examine entire cell populations, which can hide important distinctions between individual cells. Single-cell RNA sequencing (scRNA-seq) and single-cell proteomics are emerging as powerful tools for investigating gene expression and protein profiles at the individual cell level, revealing cell-to-cell variability and rare cell types, and elucidating cellular responses to environmental changes or treatments (De Mandal et al., 2022). For example, scRNA-seq has proven critical in identifying various cancer cell subpopulations and understanding their involvement in tumor growth and resistance to therapy. The continuous development of single-cell technology is expected to improve our ability to understand complicated biological systems and tailor medicinal therapies to individual cellular characteristics.

Systems biology, an integrative approach combining cell biochemistry with computational modeling, is another key trend. By constructing comprehensive models of cellular networks, researchers can simulate and predict cellular responses to various stimuli or perturbations (Murray et al., 2016). Advances in computational power and algorithms are improving the integration of multi-omics data, such as genomics, proteomics, and metabolomics, to build more accurate models of cellular function. These integrative approaches are essential for understanding the interplay between different biochemical pathways and their impact on cellular behavior. Future research in systems biology is likely to focus on refining these models, incorporating dynamic interactions, and applying them to drug discovery, disease modeling, and personalized medicine.

Synthetic biology and cellular engineering are rapidly advancing fields with significant implications for cell biochemistry research. Synthetic biology involves designing and constructing new biological parts, devices, and systems, or redesigning existing ones to achieve specific functions. Researchers are developing engineered cells with synthetic circuits that can perform complex tasks, such as detecting disease markers or producing therapeutic compounds (Mao et al., 2021). Advances in gene editing technologies, such as CRISPR/Cas9, are enabling precise modification of cellular genomes, allowing researchers to investigate gene function and create model systems for studying diseases. Future trends in synthetic biology are likely to include the development of more sophisticated cellular constructs and the application of these technologies to regenerative medicine, bioengineering, and therapeutic interventions (Murray et al., 2016).

Developments in metabolomics and the thorough investigation of metabolites in biological systems, are becoming more and more significant in the field of cell biochemistry research. The sensitivity and resolution of metabolomic investigations have been improved by advancements in mass spectrometry and nuclear magnetic resonance (NMR) spectroscopy. This has made it possible to identify a wider variety of metabolites and their dynamic changes in response to physiological or pathological situations. Changes in metabolic pathways have been linked to cancer, cardiovascular disease, and neurological illnesses. Metabolomics offers insights into cellular metabolism, disease processes, and therapeutic targets. To clarify intricate metabolic networks and their roles in health and illness, future metabolomics research is anticipated to concentrate on integrating metabolomic data with other omics technologies and creating more reliable analytical methods (De Mandal et al., 2022).

Artificial intelligence (AI) and machine learning (ML) are increasingly applied to cell biochemistry research to analyze complex datasets and uncover patterns that may not be apparent through traditional methods. AI algorithms can process large-scale data from high-throughput experiments, such as omics analyses, to identify novel biomarkers, predict drug responses, and elucidate disease mechanisms. Machine learning models are used to predict protein structures, analyze gene expression data, and model biochemical interactions, and the continued development of AI and ML is likely to accelerate discoveries and enhance the translation of research findings into clinical practice (Mao et al., 2021).

The trend toward personalized and precision medicine is closely linked to advances in cell biochemistry. By integrating genomic, proteomic, and metabolomic data, researchers can tailor treatments to individual patients based on their unique biochemical profiles. This approach aims to improve treatment efficacy and safety by targeting specific molecular mechanisms involved in a patient's disease. For example, precision oncology uses molecular profiling of tumors to identify actionable genetic alterations and guide targeted therapies. Future research in personalized medicine is expected to focus on developing comprehensive biomarkers, improving data integration techniques, and optimizing treatment strategies based on individual patient profiles (Hewitt, 2004).

Future developments in cell biochemistry research are expected to revolutionize our understanding of cellular processes and their consequences for health and disease. Advances in imaging technologies, single-cell omics, systems biology, synthetic biology, metabolomics, AI, and customized medicine are driving this transformation. As these technologies advance, they will provide deeper insights into biological systems, allow for more precise diagnostic and therapeutic methods, and eventually improve our ability to solve complicated medical difficulties (Mao et al., 2021). The combination of these developing ideas has the potential for enormous advances in both basic research and clinical applications, ushering in a new era of customized and precision medicine.

Chapter 2

Macromolecules in Cells

Contents

Cells, the basic units of life, are complex assemblies of various macromolecules, each with a critical role in maintaining cellular function, structure, and regulation. Proteins, nucleic acids, carbohydrates, and lipids are the four major types of macromolecules required for life's biochemical processes (Alberts et al., 2015). Proteins are made up of amino acids and act as enzymes, structural components, and signaling molecules, facilitating a wide range of biochemical reactions and maintaining cellular architecture.

Nucleic acids, including DNA and RNA, store and transmit genetic information, directing protein synthesis and ensuring the continuity of biological information over generations. Carbohydrates, which range from simple sugars to complex polysaccharides, primarily serve as energy sources and structural components, but they also play roles in cell recognition and signaling (Pollard et al., 2022).

Lipids, which are hydrophobic in nature, play important roles in the formation of cellular membranes, energy storage, and signaling. The intricate interplay of these macromolecules underpins cellular metabolism, communication, and replication, emphasizing their critical roles in cellular homeostasis and life perpetuation. Understanding the biochemical properties and functions of these macromolecules is critical for understanding the molecular mechanisms that control cellular behavior and other physiological processes in living organisms (Whitford, 2013).

2.1. PROTEINS: BUILDING BLOCKS OF LIFE

Proteins, also known as the building blocks of life, play an important role in almost every biological process that occurs within cells. Proteins, as polymers made up of amino acids, have a wide range of structures and functions that support the biochemical and physiological activities required for life. Proteins' intricate architecture, determined by the amino acid sequence in their polypeptide chains, serves as the foundation for their diverse functional repertoire. Each amino acid sequence is encoded by the organism's genetic material, demonstrating the central dogma of molecular biology: DNA is transcribed into RNA, which is then translated into proteins. This information transfer from nucleic acids to proteins is essential for the regulation and execution of cellular functions (Pollard et al., 2022).

Proteins have a hierarchical structure, with four levels: primary, secondary, tertiary, and quaternary. The primary structure is a linear sequence of amino acids linked by peptide bonds. This sequence is critical because it determines the protein's final conformation and, therefore, its function (Whitford, 2013). The secondary structure is formed by hydrogen bonding between the amino acid backbone atoms, resulting in the formation of alpha-helices and beta-pleated sheets that give the protein stability and shape. Tertiary structure refers to the overall three-dimensional folding of a single polypeptide chain, which is supported by a variety of interactions including hydrophobic interactions, hydrogen bonds, ionic bonds, and disulfide bridges. This level of structure is critical for the functional conformation of the protein. Some proteins have a quaternary structure, which is the assembly of multiple polypeptide chains, or subunits, into a functional complex (Alberts et al., 2015). Hemoglobin, for example, is a quaternary protein made up of four subunits, each of which contributes to the ability to transport oxygen in the blood (Figure 2.1).

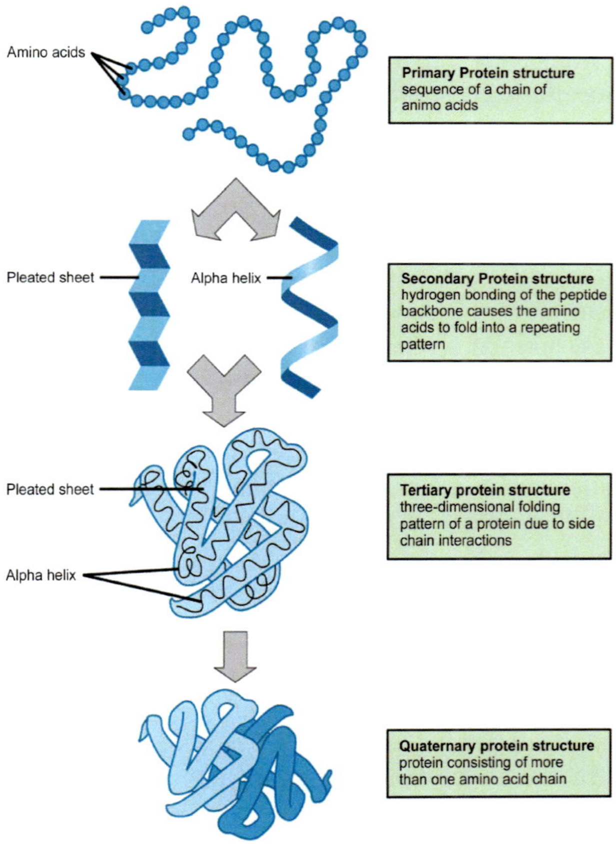

Figure 2.1. *The hierarchical structure of proteins.*

Source: https://courses.lumenlearning.com/wm-biology1/chapter/reading-protein-structure/.

Proteins serve a variety of functions within the cell, many of which are essential to life. Enzymes, a type of protein, serve as catalysts, accelerating biochemical reactions by lowering the activation energy required. This catalytic activity is critical for sustaining life because it allows metabolic reactions to occur at rates that are appropriate for cellular needs. Enzymes are highly specific, frequently binding to their substrates with exquisite precision due to the complementarity of the enzyme's active site and the substrate's structure. Enzymatic activity regulation is an important aspect of cellular control because it ensures that metabolic pathways are finely tuned to the needs of the cell (Alberts et al., 2015).

Beyond their enzymatic functions, proteins help to maintain the structural integrity of cells and tissues. Cytoskeletal proteins, such as actin and tubulin, form the cell's framework, providing shape and mechanical support while also facilitating intracellular transport and cell division. These proteins are dynamic, assembling and disassembling in response to cellular signals, allowing the cell to adapt to changing environments. Extracellularly, proteins like collagen and elastin provide strength and elasticity to connective tissues, highlighting proteins' diverse structural roles in multicellular organisms (Boal 2012).

Proteins play important roles in cellular signaling and communication. Receptor proteins are embedded in the cell membrane and detect extracellular signals such as hormones, neurotransmitters, and growth factors. When these receptors bind to their ligands, they undergo conformational changes that trigger intracellular signaling cascades, eventually leading to specific cellular responses. This process, known as signal transduction, is required to coordinate cellular activities and ensure that cells respond appropriately to their surroundings (Pollard et al., 2022). Transcription factors are another type of protein that regulates gene expression by binding to specific DNA sequences and controlling gene transcription in response to various stimuli.

Furthermore, proteins play an important role in the immune system, where antibodies, also known as immunoglobulins, recognize and neutralize foreign pathogens like viruses and bacteria. The immune system can target a wide range of invaders thanks to antibodies' specificity, which stems from the vast diversity of amino acid sequences in their antigen-binding sites. Proteins also play a role in transport, with hemoglobin serving as a prime example, transporting oxygen from the lungs to tissues while returning carbon dioxide for exhalation. Membrane transport proteins, which include channels and carriers, control the movement of ions, nutrients, and waste products across cellular membranes, thereby maintaining homeostasis (Alberts et al., 2015; Whitford, 2013). The biological importance of proteins cannot be overstated. They are involved in almost every aspect of cellular life, from the most fundamental functions like metabolism and energy production to more complex ones like cell division, differentiation, and intercellular communication. The variety of protein functions demonstrates the versatility of their molecular structure, which allows for precise interactions with a wide range of molecules, enabling the intricate regulation and coordination required for life (Boal 2012).

2.2. NUCLEIC ACIDS: GENETIC INFORMATION

Nucleic acids are fundamental macromolecules that store and transmit genetic information within cells. The two primary types of nucleic acids, deoxyribonucleic acid (DNA) and ribonucleic acid (RNA), are essential to life's continuity, guiding protein synthesis and ensuring the fidelity of genetic information across generations. These molecules are made up of monomeric units known as nucleotides, each containing a phosphate group, a pentose sugar, and a nitrogenous base. The sequence of these nucleotides encodes the genetic instructions required for all living organisms' development, function, and reproduction (Baynes and Dominiczak, 2009; Alberts et al., 2015; Pollard et al., 2022) (Figure 2.2).

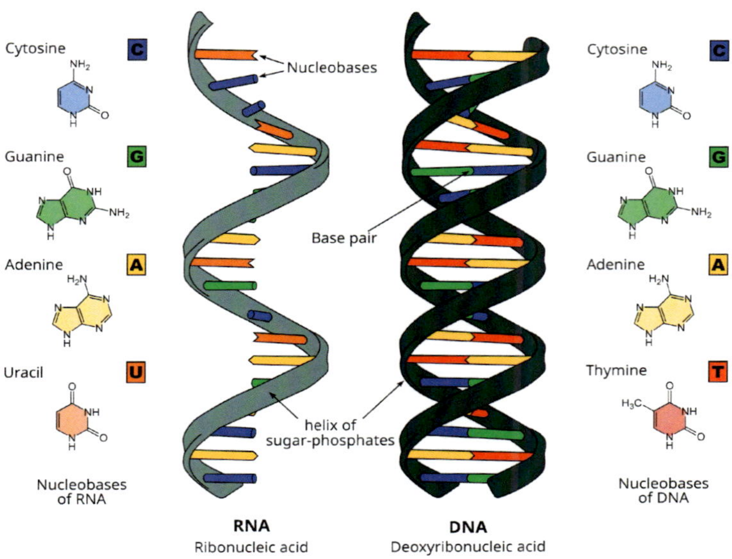

Figure 2.2. *DNA and RNA as nucleic acids.*

*Source: By File: Difference DNA RNA-DE.svg: Sponk / *translation: Sponk – Chemical structures of nucleobases by Roland1952, CC BY-SA 3.0, https://commons.wikimedia.org/w/index.php?curid=9810855.*

DNA is the principal molecule responsible for the storage of genetic information. Structurally, DNA is characterized by its double helix formation, in which two complementary strands coil around each other, stabilized by hydrogen bonds between paired nitrogenous bases. The sugar in DNA is deoxyribose, and the bases include adenine (A), thymine (T), cytosine (C), and guanine (G). The specific pairing between adenine and thymine, and between cytosine and guanine, is crucial for the accurate replication of genetic material. The double-stranded nature of DNA not only provides a stable framework for storing genetic information but also facilitates its replication during cell division (Saenger, 2013). Each strand acts as a template for the synthesis of a new complementary strand, ensuring that genetic information is transmitted correctly to daughter cells. This replication process is semi-conservative, which means that each new DNA molecule contains one original strand and one newly synthesized strand, ensuring the genetic code's integrity.

The function of DNA extends beyond information storage; it also regulates gene expression via a variety of mechanisms. DNA is packaged into chromatin, a complex of DNA and histone proteins that plays an important role in gene regulation. The degree of chromatin compaction influences DNA accessibility to transcriptional machinery, thereby regulating which genes are expressed at any given time. Epigenetic modifications, such as DNA methylation and histone acetylation, influence chromatin structure and gene expression, allowing cells to respond dynamically to environmental cues and maintain cellular identity throughout development (Csordás et al., 2006; Pollard et al., 2022).

While structurally similar to DNA, RNA differs in several key ways that allow it to perform unique functions within the cell. RNA is typically single-stranded and contains ribose as its sugar, with uracil (U) replacing thymine as a nitrogenous base. RNA is essential for the transmission and execution of genetic information via processes known as gene expression. The first step in this process is transcription, which involves synthesizing RNA from a DNA template. During transcription, RNA polymerase reads the DNA template strand and creates a complementary RNA strand, which is then processed to form a mature messenger RNA (mRNA) (Alberts et al., 2015).

mRNA is the intermediary that transports genetic information from the nucleus to the cytoplasm, where it controls protein synthesis. This process, known as translation, occurs on ribosomes, which are complex molecular machines made up of ribosomal RNA (rRNA) and proteins. The mRNA sequence is read in sets of three nucleotides called codons, each of which specifies an amino acid. Transfer RNA (tRNA) molecules are small RNA molecules with a specific structure that match the codons on the mRNA to the corresponding amino acids, allowing them to be incorporated into the growing polypeptide chain (Saenger,

2013). This precise translation of genetic information into functional proteins is critical to cellular activity, as proteins perform a wide range of functions required for life.

Besides its role in protein synthesis, RNA plays a variety of other cellular functions. The ribosome's structural integrity and catalytic activity are dependent on rRNA, which is its primary component. In addition to its role in translation, tRNA helps to ensure the correct amino acids are incorporated during protein synthesis. Small nuclear RNA (snRNA) and small nucleolar RNA (snoRNA) are involved in the processing of pre-mRNA and rRNA, demonstrating RNA's versatility in cellular metabolism (Pollard et al., 2022).

Furthermore, non-coding RNAs, which do not encode proteins, have emerged as important regulators of gene expression. MicroRNAs (miRNAs) and small interfering RNAs (siRNAs) play a role in post-transcriptional regulation by binding to target mRNAs and either degrading or inhibiting their translation, thereby fine-tuning protein levels within cells. Long non-coding RNAs (lncRNAs) have a variety of functions, including chromatin remodeling, transcriptional regulation, and genome integrity. These regulatory RNAs add complexity to gene expression, allowing for precise control of cellular function and adaptation to environmental changes (Saenger, 2013; Alberts et al., 2015).

The interaction of DNA and RNA is critical to the transmission of genetic information within cells. The stability of DNA ensures the preservation of the genetic blueprint, whereas the versatility of RNA allows for dynamic gene expression and regulatory functions. Mutations or changes in the sequence or structure of nucleic acids can have a significant impact on cellular function and organismal health, resulting in a variety of genetic disorders and illnesses. Understanding the molecular mechanisms governing nucleic acid function is thus critical for developing therapeutic strategies aimed at correcting genetic defects and treating diseases at their source (Whitford, 2013).

2.3. CARBOHYDRATES: ENERGY SOURCE

Carbohydrates perform a variety of functions that are critical to the survival and proper functioning of living organisms. Among their many functions, carbohydrates are best known as the primary energy source for cellular activities. These organic compounds, made up of carbon, hydrogen, and oxygen, are divided into three types: monosaccharides, disaccharides, and polysaccharides (Pigman 2012) (Figure 2.3).

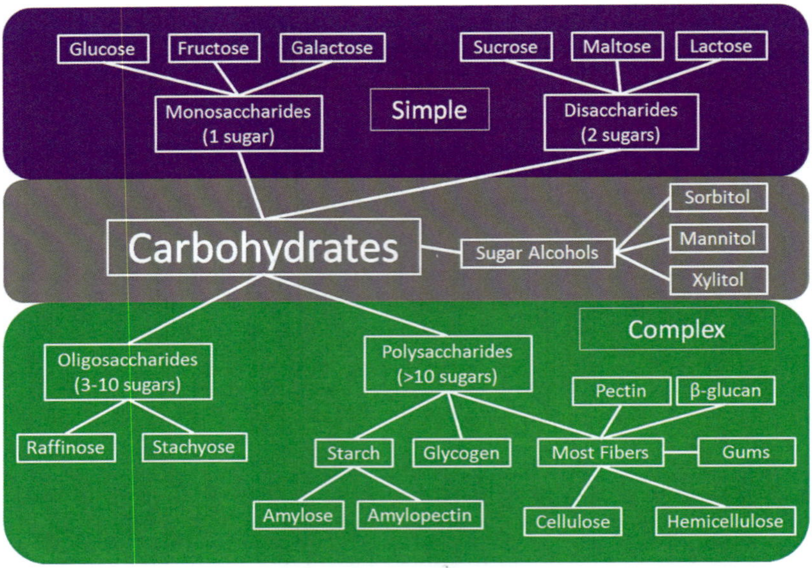

Figure 2.3. *Carbohydrate classification scheme.*

Source: https://courses.lumenlearning.com/suny-nutrition/chapter/2-11-simple-carbohydrates/.

Monosaccharides, the simplest type of carbohydrate, serve as the foundation for the creation of more complex carbohydrates. Glucose, a six-carbon monosaccharide, is the most common and biologically important. It is the primary substrate for cellular respiration, a series of metabolic pathways that convert glucose into the cell's universal energy currency, adenosine triphosphate (ATP). Glycolysis, the first stage of glucose metabolism, converts glucose into two molecules of pyruvate, producing a small amount of ATP and reducing equivalents in the form of nicotinamide adenine dinucleotide (NADH). The generated pyruvate can then enter the mitochondria, where it undergoes further oxidation in the citric acid cycle and the electron transport chain, resulting in the production of a significantly larger amount of ATP (Pigman, 2012; Alberts et al., 2015). This efficient conversion of glucose into energy underscores its central role in maintaining cellular functions.

Disaccharides, such as sucrose and lactose, are made up of two monosaccharide units connected by glycosidic bonds. These molecules are common in the diet and serve as energy sources, which are easily broken down into monosaccharides by digestive enzymes. The monosaccharides are then absorbed into the bloodstream and transported to cells, where they can be used in metabolic processes to produce ATP. Disaccharides' ability to hydrolyze quickly into monosaccharides makes them an important source of rapid energy, especially in organisms that require immediate energy access, such as muscle cells during intense physical activity (Boal, 2012).

Polysaccharides, the most complex type of carbohydrate, are long chains of monosaccharide units held together by glycosidic bonds. These macromolecules primarily serve as energy storage and structural components of cells. Glycogen, a highly branched polysaccharide of glucose, is the primary storage form of carbohydrates in animals. It is primarily stored in the liver and muscle tissues, where it can be quickly mobilized to release glucose as energy demands rise. The process of glycogenolysis, which breaks down glycogen into glucose, ensures a consistent supply of glucose for ATP production, particularly during periods of fasting or intense exercise (Baynes and Dominiczak, 2009). This storage mechanism is vital for maintaining blood glucose levels and providing a quick source of energy to meet the body's metabolic needs.

Polysaccharides are important structural elements in cells in addition to serving as energy reserves. Cellulose, a polysaccharide made up of β-glucose units, is a structural component of plant cell walls that provides rigidity and resistance to mechanical stresses. Strong hydrogen bonds between β-glucose units in cellulose chains create a stable and insoluble fiber, essential for plant cell structural integrity. In animals, glycosaminoglycans, complex polysaccharides that contribute to the extracellular matrix and provide support and protection to tissues, fulfill similar structural roles (Boal, 2012; Alberts et al., 2015).

Carbohydrates are also involved in cell signaling and recognition processes, which help to regulate a variety of cellular functions. Glycoproteins and glycolipids are carbohydrates that are covalently attached to proteins and lipids, respectively, and can be found on cell membranes. These carbohydrate moieties serve as recognition sites for specific interactions between cells and their surroundings, with important roles in immune responses, cell-cell communication, and pathogen recognition (Baynes and Dominiczak, 2009; Pigman, 2012). For example, ABO blood group antigens are glycoproteins found on the surface of red blood cells that determine an individual's blood type based on the specific carbohydrate structures they contain. The precise carbohydrate patterns on glycoproteins and glycolipids can influence cellular interactions and play an important role in processes such as fertilization, development, and immunity.

Furthermore, carbohydrates are necessary for the synthesis of nucleotides, which are the building blocks of nucleic acids. Ribose and deoxyribose, five-carbon sugars, are found in RNA and DNA, respectively. These sugars are required for the formation of nucleic acid backbones, which link with phosphate groups and nitrogenous bases to form the molecules' structural frameworks

(Pigman, 2012; Saenger, 2013). Carbohydrates' involvement in nucleotide synthesis emphasizes their importance beyond energy metabolism, as they are also required for the storage and transmission of genetic information.

Carbohydrates play an important role in cellular metabolism, including metabolic pathways other than glycolysis. For example, the pentose phosphate pathway, which runs parallel to glycolysis, uses glucose-6-phosphate to produce ribose-5-phosphate for nucleotide synthesis and NADPH for reductive biosynthesis. NADPH is especially important in the defense against oxidative stress because it provides reducing power for the regeneration of glutathione, an antioxidant that protects cells from reactive oxygen species (Pigman 2012).

2.4. LIPIDS: CELL MEMBRANE STRUCTURE

Lipids are a diverse group of hydrophobic macromolecules that play important roles in cellular structure, energy storage, and signaling. Among their many functions, lipids are best known as the primary components of cell membranes, where they serve as the fundamental barrier that defines cellular boundaries and mediates interactions with the external environment. Their unique biochemical properties, particularly their amphipathic nature, underpin their ability to assemble into membrane bilayer structures, which is critical for cellular integrity and function (van Meer et al., 2008).

The lipid bilayer, a dynamic and flexible matrix composed primarily of phospholipids, serves as the structural foundation for cell membranes. Phospholipids are amphipathic molecules, which means they have hydrophilic (water-attracting) and hydrophobic (water-repelling) regions (Baynes and Dominiczak, 2009). Each phospholipid is made up of a glycerol backbone, two fatty acid chains, and a phosphate group. The hydrophobic fatty acid chains align away from water, whereas the hydrophilic phosphate-containing head group interacts with the water. This dual nature causes the phospholipids to spontaneously form a bilayer in an aqueous environment, with the hydrophobic tails facing inward and shielded from water, and the hydrophilic heads facing outward and interacting with the surrounding water. This arrangement creates a semi-permeable membrane that acts as a selective barrier, regulating the passage of ions, molecules, and signals into and out of the cell (Feingold, 2024).

The lipid bilayer is a fluid structure made up of lipids and proteins that move laterally. This fluidity is critical for membrane function because it allows proteins and lipids to diffuse within the bilayer, facilitating cellular processes like signal transduction, membrane trafficking, and the formation of specialized membrane domains known as lipid rafts. Lipid rafts are microdomains rich in cholesterol, sphingolipids, and specific proteins that act as organizing centers for the assembly of signaling molecules, influencing membrane fluidity and protein function. The fluid mosaic model of membrane structure, first proposed by Singer and Nicolson in 1972, emphasizes this dynamic nature of membranes, where proteins and lipids coexist in a fluid matrix, allowing for a wide range of cellular activities (van Meer et al., 2008).

Cholesterol, another important lipid in cell membranes, regulates membrane fluidity and stability. Cholesterol intercalates between phospholipid molecules, and its rigid steroid ring structure reduces membrane fluidity at high temperatures while preventing membrane crystallization at low temperatures. This bidirectional regulatory role keeps the membrane fluid and functional at a variety of physiological temperatures. Cholesterol also aids in the formation of lipid rafts, which influences membrane organization and signaling (van Meer et al., 2008).

The specific fatty acid composition of phospholipids influences membrane properties. Saturated fatty acids, which lack double bonds, pack more tightly together, resulting in a more rigid membrane, whereas unsaturated fatty acids, which contain one or more double bonds, cause kinks in the fatty acid chains, preventing tight packing and increasing membrane fluidity. Cells tightly regulate the balance of saturated and unsaturated fatty acids in membrane phospholipids to ensure optimal membrane fluidity, which is required for proper membrane protein function and cellular processes (Feingold, 2024).

Beyond their structural role, lipids in cell membranes are actively involved in signaling processes. Certain phospholipids, such as phosphatidylinositol, can be phosphorylated to generate phosphoinositides, which serve as docking sites for signaling proteins and play key roles in pathways regulating cell growth, survival, and metabolism. Moreover, the hydrolysis of membrane phospholipids by enzymes such as phospholipases generates second messengers like diacylglycerol (DAG) and inositol triphosphate (IP3), which propagate intracellular signals that lead to various cellular responses, including changes in gene expression, calcium mobilization, and cytoskeletal rearrangements (Baynes and Dominiczak, 2009).

Sphingolipids, another class of membrane lipids, also contribute to the structural and functional complexity of cell membranes. These lipids, which include sphingomyelin and glycosphingolipids, are prominent components of the outer leaflet of the plasma membrane. Sphingolipids are involved in cell-cell recognition, signal transduction, and apoptosis. Glycosphingolipids, which have sugar moieties attached to their lipid backbone, are particularly important in the formation of membrane microdomains and in mediating interactions with the extracellular matrix and other cells (Boal, 2012; Feingold, 2024). The diverse functions of sphingolipids underscore the versatility of lipids in cellular membranes, extending beyond structural roles to include critical regulatory functions.

Lipids also act as precursors for bioactive molecules like eicosanoids, which are formed from arachidonic acid, a polyunsaturated fatty acid found in membrane phospholipids. Eicosanoids, which include prostaglandins, thromboxanes, and leukotrienes, are strong signaling molecules that control inflammation, immune responses, and vascular function. The release of arachidonic acid from membrane phospholipids by phospholipase A2 is an important step in the biosynthesis of eicosanoids, connecting membrane lipid composition to the regulation of physiological processes and pathophysiological conditions (Alberts et al., 2015).

Membrane structural integrity is also important for the formation and maintenance of cellular organelles, which are surrounded by lipid membranes that define their boundaries and compartmentalize cellular functions. For example, the mitochondrial membrane, which is high in cardiolipin, is required for the organization of the electron transport chain and the production of ATP via oxidative phosphorylation. The mitochondrial membrane's unique lipid composition facilitates its role in energy production, emphasizing membrane lipid specialization in various cellular contexts (Pollard et al., 2022; Feingold, 2024).

2.5. ENZYMES: BIOLOGICAL CATALYSTS

Enzymes are pivotal to life, serving as biological catalysts that accelerate chemical reactions in living organisms. These proteins, highly specialized for their respective tasks, increase the rate of biochemical reactions without undergoing permanent change themselves. The mechanism by which enzymes function within cells is critical to the metabolic activities that support life. From digestion to DNA replication, enzymes help to convert substrates into products with extraordinary selectivity and efficiency. Understanding enzyme structure, function, and kinetics reveals their involvement in cellular homeostasis. Enzymes are made up of amino acids that fold into three-dimensional structures, resulting in distinct active sites adapted to certain substrates. Binding a substrate to the enzyme's active site creates an enzyme-substrate complex, which stabilizes the transition state and lowers the activation energy of the activity. Enzyme specificity is sometimes compared to a lock and key mechanism, in which only a certain substrate (the key) fits into the enzyme's active region (the lock) (Palmer and Bonner, 2007). Enzymes have extraordinary specificity due to the precise arrangement of amino acids at the active site, which can establish hydrogen bonds, hydrophobic interactions, and ionic connections with the substrate. Some enzymes display induced fit, which occurs when the substrate binds to the enzyme, causing a conformational change that optimizes its interaction with the substrate and improves catalysis (Pollard et al., 2022).

Enzymes speed up biological reactions by stabilizing the transition state and lowering the activation energy. Several processes contribute to the catalytic action. One frequent technique is proximity and orientation, in which the enzyme gets substrates close together and aligns them in the proper orientation for the reaction to occur. This reduces the reaction's entropy barrier and raises the possibility of productive substrate collisions. Another mechanism is acid-base catalysis, in which amino acid residues in the enzyme's active site act as proton donors or acceptors, allowing proton transfer throughout the reaction. For example, the enzyme ribonuclease A employs histidine residues to contribute and absorb protons during the hydrolysis of the RNA (Punekar, 2018). Enzymes can also use covalent catalysis, which occurs when an enzyme and a substrate form a brief covalent link. This mechanism is demonstrated by serine proteases, which produce a covalent intermediate during peptide bond hydrolysis. Furthermore, metal ion catalysis requires the presence of metal cofactors such as zinc or magnesium, which stabilize negative charges or enhance electron transfer during the process (Punekar, 2018).

The study of enzyme kinetics sheds light on the speeds of biological reactions and how they are affected by variables such as substrate concentration, temperature, and pH. The Michaelis-Menten equation expresses the relationship between the reaction rate (v) and the substrate concentration ([S]). This equation includes two important parameters: the maximal reaction rate (Vmax) and the Michaelis constant (Km), which represents the substrate concentration at half the maximum rate. The kinetics of enzymes are critical to understanding their efficiency and regulation in cellular metabolism. Some enzymes have simple Michaelis-Menten kinetics, while others show more sophisticated behavior, such as cooperative binding or allosteric control. In cooperative enzymes, the binding of a substrate to one active site enhances the binding of substrates to other active sites, leading to a sigmoidal relationship between reaction rate and substrate concentration (Pollard et al., 2024).

Cells have strict regulation of enzyme activity to guarantee that metabolic pathways function well and in accordance with biological requirements. One way of control is allostery, in which an effector molecule bound to a location other than the active site causes the enzyme to alter its shape. This may increase or decrease the activity of the enzyme. For instance, when energy levels are adequate, high amounts of ATP allosterically inhibit phosphofructokinase, a crucial enzyme in glycolysis, avoiding excessive glucose breakdown. Another typical regulation mechanism in metabolic pathways is feedback inhibition. In this process, the end product of a pathway inhibits an earlier enzyme in the pathway, preventing the accumulation of intermediates and conserving cellular resources. For example, the amino acid isoleucine inhibits the first enzyme in its biosynthetic pathway, threonine deaminase, ensuring that isoleucine synthesis is tightly regulated (Punekar, 2018).

Enzymes are essential for metabolism because they catalyze the synthesis of biomolecules required for cell growth and maintenance as well as the breakdown of nutrients to produce energy. For cellular respiration, enzymes like ATP synthase are necessary because they enable the mitochondria to produce ATP from ADP and inorganic phosphate. Enzymes that are involved in the citric acid cycle, glycolysis, and oxidative phosphorylation are also essential for the production of energy (Keller et al., 2015). Metabolic abnormalities and disease can result from defects in the action of enzymes. For instance, phenylketonuria (PKU) is brought on by a phenylalanine hydroxylase enzyme deficiency, which builds up dangerously high blood phenylalanine levels. Emerging therapies for such genetic enzyme deficits include gene therapy and enzyme replacement therapy (Pollard et al., 2024). Enzymes are also targets for drug development, with many pharmaceuticals designed to inhibit specific enzymes involved in disease processes. For example, statins inhibit the enzyme HMG-CoA reductase, reducing cholesterol synthesis and lowering the risk of cardiovascular disease.

In several industrial and biotechnological applications, the special qualities of enzymes have been utilized. Biofuels, medications, and food items are all produced with the aid of enzymes. Chemical catalysts frequently require harsh conditions and produce undesired byproducts; in contrast, biological catalysts can catalyze reactions under mild circumstances with excellent selectivity (Keller et al., 2015). Such catalysts are interesting alternatives to chemical catalysts. Industrial operations frequently use immobilized enzymes, which are enzymes that are affixed to a solid support. Enzyme reusability and easier product extraction from the reaction mixture are made possible by this. It is possible to produce bioethanol, detergents, and high-fructose corn syrup using immobilized enzymes (Pollard et al., 2022).

Enzymes are essential biological catalysts that accelerate metabolic reactions with high specificity and efficiency. Their involvement in metabolism, illness, and industrial applications emphasizes their significance in biology and biotechnology. Advances in our understanding of enzyme structure, activity, and regulation continue to drive medical and industrial innovation, with enzymes functioning as adaptable tools for a variety of applications.

2.6. REGULATION OF MACROMOLECULE SYNTHESIS

In cell biochemistry, macromolecule synthesis regulation is a critical mechanism for cell growth, division, and environmental adaptability. To maintain cellular integrity and function, macromolecules such as proteins, nucleic acids, carbohydrates, and lipids are created in closely controlled conditions. Cells' regulatory systems consist of intricate metabolic processes that are controlled by enzymes, regulatory proteins, and signaling molecules. Understanding these regulatory pathways provides information on cell physiology, disease causes, and possible therapeutic approaches (Figure 2.4).

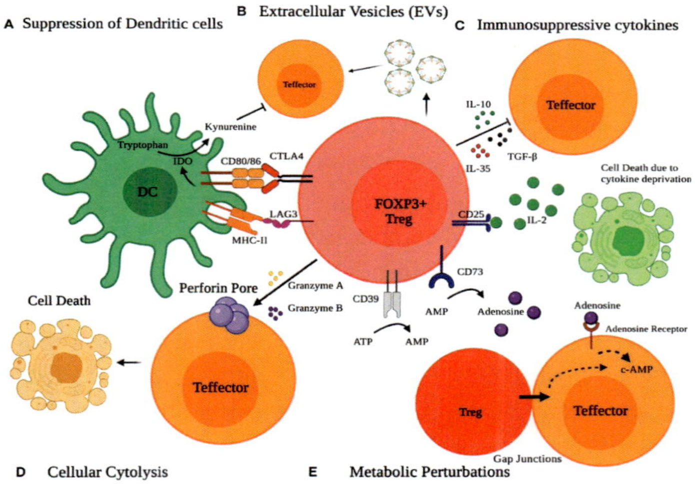

Figure 2.4. *How regulatory T cells work.*

Source: Grover P, Goel PN and Greene MI (2021) Regulatory T Cells: Regulation of Identity and Function. Front. Immunol. 12:750542. doi: 10.3389/fimmu.2021.750542.

A variety of methods are employed by cells to regulate the synthesis of macromolecules, guaranteeing that these processes are following the cell's metabolic requirements and external milieu. The control of RNA synthesis and ribosome biogenesis is a crucial component of this regulation. For protein synthesis to occur, ribosome production needs to be strictly regulated in response to the cell's energy and nutrition resources (Punekar, 2018). One example of how cells strike a balance between growth and metabolism is the regulation of ribosome synthesis.

The regulation of enzyme activity is one of the basic regulatory points in the production of macromolecules. The regulation of enzymes involved in the synthesis of proteins, lipids, and nucleic acids is achieved by feedback inhibition, covalent modifications, and allosteric interactions. For example, by modifying the phosphorylation state of target proteins, protein kinase C (PKC) is essential for controlling several biological functions, such as the synthesis of macromolecules (Punekar, 2018). PKC can modify the function of proteins that are involved in the creation of macromolecules, allowing extracellular signals to be integrated with intracellular metabolic processes.

Moreover, cells control the production of macromolecules by feedback inhibition. The control over the synthesis of cholesterol is a prime example. The production of cholesterol is a multi-step process that is regulated by enzyme activity and gene expression, among other factors. When cellular levels are enough, cholesterol itself functions as a feedback inhibitor, reducing its own synthesis (Alberts et al., 2015). This guarantees that resources aren't squandered on making extra cholesterol, which might be harmful to the cells.

Ferritin synthesis is another example of macromolecular control, specifically in the setting of iron homeostasis. Ferritin is a protein complex that accumulates and releases iron in a controlled manner to maintain normal cellular iron levels. Iron-responsive elements (IREs) in mRNA regulate ferritin synthesis through interactions with iron regulatory proteins (IRPs). This connection regulates whether ferritin mRNA is translated into protein in response to the cell's iron level (Mackiewicz et al., 2007). Thus, ferritin synthesis is controlled in response to the cell's iron requirements, limiting the accumulation of excess free iron, which could catalyze the generation of damaging reactive oxygen species.

The synthesis of macromolecules is governed not only by individual enzymes and proteins, but also by the coordination of enormous molecular complexes. Molecular chaperones, for example, aid in the appropriate folding of freshly generated proteins, ensuring that they reach the optimal shape for functional activity (Pollard et al., 2024). These chaperones regulate protein synthesis by preventing misfolded proteins from aggregating and encouraging the formation of multiprotein complexes.

Understanding how cellular signaling pathways interact with the machinery that produces macromolecules is an important topic of research in macromolecule synthesis regulation. Signaling mechanisms, such as the mammalian target of rapamycin (mTOR) pathway are critical for coordinating cell growth with macromolecule production, particularly proteins and lipids. The mTOR pathway regulates protein synthesis by modulating the activities of ribosomal protein S6 kinases and eukaryotic initiation factors (Baynes and Dominiczak, 2009). Dysregulation of the mTOR pathway is linked to a variety of illnesses, including cancer, due to its function in encouraging unregulated cell growth and proliferation.

In addition to enzymatic and signaling regulation, the spatial arrangement of macromolecular synthesis within the cell is crucial. Macromolecules like DNA and RNA are generated in distinct cellular compartments, ensuring that their production is in sync with cellular processes. For example, to achieve proper genetic replication, DNA synthesis during cell division must be precisely regulated and localized to the nucleus (Punekar, 2018). Disruptions in this spatial regulation can cause genetic instability and the onset of illnesses like cancer.

Finally, the regulation of macromolecule production goes beyond individual cells and includes intercellular signaling and communication. For example, in multicellular organisms, hormones and cytokines can control macromolecule synthesis in target cells. During immunological reactions, cytokines generated by immune cells can encourage the creation of immune-related proteins such as antibodies and cytokine receptors (Mackiewicz et al., 2007). These regulatory mechanisms guarantee that macromolecule synthesis is coordinated across tissues and organs, allowing the organism to adapt to physiological demands and environmental challenges.

2.7. ROLE OF MACROMOLECULES IN CELL SIGNALING

Macromolecules play a critical role in the regulation and functioning of cellular processes, particularly in cell signaling, which governs how cells respond to various external stimuli. The complexity of cellular signaling pathways, involving proteins, nucleic acids, lipids, and carbohydrates, is crucial for maintaining homeostasis, ensuring proper cellular communication, and regulating cellular responses to environmental cues (Alberts et al., 2002).

2.7.1. Proteins as Signaling Molecules

Proteins are the primary macromolecules involved in cell signaling. Proteins perform a variety of roles in the intricate interactions that occur within and between cells, including those of scaffolds, receptors, enzymes, and signaling molecules. G protein-coupled receptors (GPCRs), receptor tyrosine kinases, and ion channels are examples of receptor proteins that function as sensors to identify external signals including growth factors, hormones, and neurotransmitters (van Meer et al., 2008). One of the most well-known examples is the GPCR, a large family of membrane proteins responsible for detecting molecules outside the cell and activating internal signal transduction pathways. Upon ligand binding, GPCRs undergo conformational changes that activate intracellular G proteins. These proteins further propagate the signal by modulating the activity of downstream effectors, such as enzymes and ion channels, ultimately altering cellular responses like gene expression, metabolism, and cell proliferation (Pollard et al., 2024). Kinases, another type of protein macromolecule, serve critical functions in phosphorylation-based signaling cascades. Protein kinases, such as the MAPK (mitogen-activated protein kinase) pathway, phosphorylate target proteins, affecting their activity, location, and interaction with other cellular components. This phosphorylation, which is frequently reversible via the action of phosphatases, allows for precise control over cellular signaling and ensures that cellular responses can be fine-tuned and adapted to changing situations (van Meer et al., 2008).

2.7.2. Lipids in Signal Transduction

Lipids also play important roles in cell signaling, particularly within membranes where they affect signal transduction. Phosphoinositides, a kind of lipid, play important roles in many signaling cascades. For example, phosphatidylinositol (PI) and its phosphorylated derivatives (e.g., PI(4,5)P2) are essential for the activation of the PI3K-Akt pathway, which controls cell growth, survival, and (van Meer et al., 2008). In this route, PI3K (phosphoinositide 3-kinase) phosphorylates PI(4,5)P2 to become PI(3,4,5)P3, which serves as a docking site for signaling proteins with pleckstrin homology (PH) domains.

Another example is the role of sphingolipids, which regulate cellular stress responses and apoptosis. Sphingosine-1-phosphate (S1P) functions as a potent signaling molecule that controls processes like cell migration, proliferation, and survival by interacting with a family of specific GPCRs known as S1P receptors (Feingold, 2024). Lipids also play a role in cell signaling as secondary messengers. Phospholipase C (PLC) cleaves phosphatidylinositol bisphosphate (PIP2) to generate diacylglycerol (DAG) and inositol triphosphate (IP3). DAG remains within the plasma membrane and activates protein kinase C (PKC), whereas IP3 diffuses into the cytoplasm and causes the endoplasmic reticulum to release calcium ions. These second messengers amplify the signal and regulate a variety of downstream cellular responses, such as enzyme activity, gene expression, and cell shape (Punekar, 2018).

2.7.3. Nucleic Acids and Cell Signaling

Nucleic acids, particularly in the form of RNA, have emerged as important participants in cell signaling. Aside from their roles in genetic information transfer, RNAs, particularly non-coding RNAs like microRNAs (miRNAs) and long non-coding RNAs (lncRNAs), are thought to regulate gene expression and cellular signaling pathways. miRNAs work by binding to complementary regions in messenger RNAs (mRNAs), which causes mRNA degradation or translation inhibition. MiRNAs influence the expression of genes involved in cell proliferation, differentiation, and death via this method, and they are frequently misregulated in illnesses such as cancer (Pollard et al., 2024).

Moreover, circular RNAs (circRNAs), a class of non-coding RNAs, have recently been shown to regulate signal transduction by acting as molecular sponges that sequester miRNAs, thereby preventing them from inhibiting their target mRNAs (Pollard et al., 2024). This regulation highlights the diverse and intricate ways in which nucleic acids contribute to the fine-tuning of cell signaling networks.

2.7.4. Carbohydrates and Cell Signaling

Carbohydrates, specifically glycoproteins and glycolipids, also play a role in cell signaling. These macromolecules are present on the cell surface and are required for cell-to-cell communication, immunological identification, and environmental response (Punekar, 2018). Glycosylation, or the binding of carbohydrates to proteins or lipids, is an important post-translational modification that influences protein folding, stability, and function.

Lectins are carbohydrate-binding proteins that detect and bind to specific carbohydrate structures on cell surfaces, thereby mediating cell adhesion and signaling. Lectin attachment to glycosylated receptors can activate intracellular signaling pathways that regulate cell motility, immunological responses, and inflammation (Feingold, 2024).

The role of carbohydrates in cell signaling is further exemplified by the selectins, a family of cell adhesion molecules that mediate the interaction between leukocytes and endothelial cells during inflammation. Selectins bind to specific carbohydrate motifs on the surfaces of leukocytes, facilitating their rolling and subsequent extravasation into tissues at sites of injury or infection (Han et al., 2018).

Chapter 3

Metabolic Pathways

Contents

Metabolic pathways are essential to cellular biochemistry, representing a series of enzyme-catalyzed reactions that convert substrates into products required for cellular function. These pathways are complexly organized into anabolic and catabolic processes. Anabolic pathways involve the synthesis of complex molecules from simpler precursors while using energy, primarily adenosine triphosphate (ATP). Catabolic pathways, on the other hand, break down complex molecules into simpler ones, releasing energy that the cell can use for a variety of functions such as homeostasis and biosynthetic processes (Bonner and Varner, 2012).

The regulation of enzymatic activity is critical to metabolic pathways because it ensures that metabolic flux responds to cellular demands. This regulation is accomplished through allosteric control, covalent modifications, and feedback inhibition, which allow cells to adapt to changing internal and external environments (Alberts et al., 2015). Glycolysis and the citric acid cycle are prime examples of metabolic pathways, which play critical roles in energy production by oxidizing glucose and other substrates.

Understanding these pathways is critical for elucidating the biochemical basis of diseases and developing therapeutic interventions. Insights from studies on metabolic networks offer a profound comprehension of cellular metabolism.

3.1. OVERVIEW OF CELLULAR RESPIRATION

Cellular respiration is the fundamental metabolic process by which cells convert biochemical energy from nutrients into adenosine triphosphate (ATP), the cell's primary energy currency. This process is essential for cellular activity and maintaining homeostasis. Cellular respiration is divided into different stages: glycolysis, pyruvate oxidation, Krebs cycle, and electron transport chain, all of which contribute to the efficient extraction and utilization of energy from organic molecules (Alberts et al., 2015) (Figure 3.1).

The first stage of cellular respiration, glycolysis, takes place in the cytoplasm and involves the enzymatic breakdown of one molecule of glucose into two molecules of pyruvate. This process results in a net gain of two ATP molecules and two reduced nicotinamide adenine dinucleotide (NADH) molecules. Glycolysis is an anaerobic pathway because it does not require oxygen. It also serves as the foundation for both aerobic and anaerobic respiration.

The pyruvate produced during glycolysis is transported into the mitochondria, where it is oxidatively decarboxylated to form acetyl-CoA and enters the citric acid cycle. The citric acid cycle is a series of enzyme-catalyzed reactions that oxidize acetyl-CoA to carbon dioxide, producing high-energy electron carriers NADH and flavin adenine dinucleotide (FADH2), as well as a small amount of ATP (Alberts et al., 2015). These electron carriers are critical in the next step of cellular respiration.

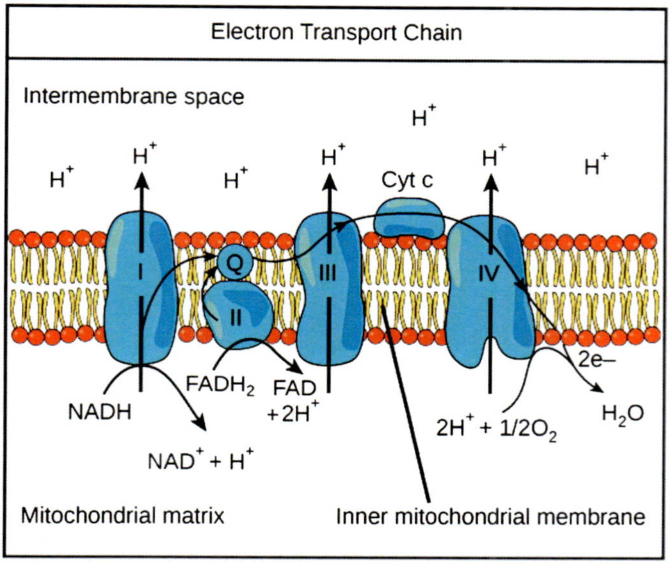

Figure 3.1. *The stages of cellular respiration.*

Source: https://courses.lumenlearning.com/suny-wmopen-biology1/chapter/cellular-respiration/.

The final stage, oxidative phosphorylation, occurs on the inner mitochondrial membrane, which contains the electron transport chain (ETC). Electrons from NADH and FADH2 are transferred through a series of protein complexes, resulting in the active transport of protons across the membrane and the formation of an electrochemical gradient. ATP synthase uses the energy stored in this gradient to phosphorylate adenosine diphosphate (ADP) to ATP, a process known as chemiosmosis. Oxygen is the terminal electron acceptor, resulting in water as a byproduct, and its presence is required for efficient ATP production (Baynes and Dominiczak, 2009).

The precise regulation of cellular respiration is critical for cellular function because it balances energy production and metabolic demands. Aberrations in this process have been linked to a variety of metabolic disorders, emphasizing the significance of understanding cellular respiration in the context of cellular biochemistry and disease.

3.2. GLYCOLYSIS AND GLUCONEOGENESIS

Glycolysis and gluconeogenesis are two essential metabolic pathways in cellular biochemistry that govern the breakdown and synthesis of glucose, respectively. These processes are crucial for maintaining energy balance and homeostasis in cells, particularly in organs like the liver, which plays a pivotal role in glucose metabolism (Alberts et al., 2015). Glycolysis is the catabolic pathway responsible for the conversion of glucose into pyruvate, yielding ATP and NADH, while gluconeogenesis is the anabolic pathway that synthesizes glucose from non-carbohydrate precursors. Despite being opposing pathways, glycolysis and gluconeogenesis share some enzymes and intermediates, but are distinct in their regulation and physiological roles (Chandel, 2021).

Glycolysis is a 10-step process that commences in the cytoplasm with the phosphorylation of glucose to glucose-6-phosphate, a process catalyzed by the enzyme hexokinase. This phosphorylation, which uses one molecule of ATP, traps glucose within the cell and commits it to further metabolism. Phosphoglucose isomerase then isomerizes glucose-6-phosphate to fructose-6-phosphate. The second phosphorylation step, catalyzed by phosphofructokinase-1 (PFK-1), converts fructose-6-phosphate to fructose-1,6-bisphosphate, another irreversible step in glycolysis that is tightly regulated and serves as a key control point for the pathway. ATP allosterically inhibits PFK-1 activity and AMP activates it, reflecting the cell's energy status and ensuring that glycolysis is upregulated when energy demand is high (Alberts et al., 2015; Chandel, 2021).

Aldolase then converts fructose-1,6-bisphosphate into two three-carbon intermediates: glyceraldehyde-3-phosphate (G3P) and dihydroxyacetone phosphate (DHAP). The action of triose phosphate isomerase allows these two molecules to be interconverted, and only G3P passes through glycolysis. Glyceraldehyde-3-phosphate dehydrogenase oxidizes and phosphorylates G3P, resulting in the production of 1,3-bisphosphoglycerate and NADH (Baynes and Dominiczak, 2009). This step is significant because it uses energy from G3P oxidation to reduce NAD^+ to NADH, which can then be used to generate ATP via oxidative phosphorylation.

In a reaction catalyzed by phosphoglycerate kinase, the high-energy compound 1,3-bisphosphoglycerate donates a phosphate group to ADP, resulting in ATP and 3-phosphoglycerate. This substrate-level phosphorylation marks the first instance ATP is generated in glycolysis. Subsequently, phosphoglycerate mutase converts 3-phosphoglycerate to 2-phosphoglycerate, which is then dehydrated by enolase to yield phosphoenolpyruvate (PEP), another high-energy intermediate. The final step of glycolysis is the transfer of a phosphate group from PEP to ADP, which produces a second molecule of ATP and pyruvate, the end product of glycolysis, in a reaction catalyzed by pyruvate kinase. The regulation of pyruvate kinase, like that of PFK-1, is critical for controlling glycolytic flux, with the enzyme allosterically activated by fructose-1,6-bisphosphate and inhibited by ATP (Chandel, 2021).

Glycolysis, as a whole, results in a net gain of two ATP molecules and two NADH molecules per molecule of glucose, and the pyruvate produced can enter the mitochondria for further oxidation in the citric acid cycle under aerobic conditions, or be converted into lactate under anaerobic conditions (Alberts et al., 2015). The fate of pyruvate is determined by the cell's oxygen availability and energy requirements. Additionally, glycolysis intermediates serve as precursors for various biosynthetic pathways, linking energy production with anabolic needs.

Gluconeogenesis, as opposed to glycolysis, is the process of producing glucose from non-carbohydrate precursors such as lactate, glycerol, and amino acids, primarily in the liver and, to a lesser extent, in the kidneys. This pathway is critical during fasting, intense exercise, or low-carbohydrate diets, when endogenous glucose production is required to maintain blood glucose levels and provide energy to glucose-dependent tissues like the brain and red blood cells (Melkonian et al., 2024).

Gluconeogenesis shares seven of the 10 enzymes involved in glycolysis, but the pathway is not simply the opposite of glycolysis because three irreversible steps in glycolysis must be bypassed in gluconeogenesis. The first of these steps is to convert pyruvate into phosphoenolpyruvate (PEP). Pyruvate is first carboxylated to oxaloacetate in the mitochondria by pyruvate carboxylase, a reaction that needs biotin as a cofactor and is activated by acetyl-CoA (Chourpiliadis and Mohiuddin, 2024). Oxaloacetate is then converted to PEP by phosphoenolpyruvate carboxykinase (PEPCK), which uses GTP as an energy source. This two-step reaction is essential for gluconeogenesis and is tightly regulated by hormones like glucagon and cortisol, which upregulate PEPCK expression during times of glucose scarcity.

The second irreversible step in glycolysis that is bypassed in gluconeogenesis is the conversion of fructose-1,6-bisphosphate to fructose-6-phosphate, catalyzed by fructose-1,6-bisphosphatase. This enzyme is allosterically inhibited by AMP and fructose-2,6-bisphosphate, which serve as indicators of the cell's energy status and ensure that gluconeogenesis is downregulated when energy is plentiful. The final irreversible step is the dephosphorylation of glucose-6-phosphate to free glucose by glucose-6-phosphatase, which is located in the endoplasmic reticulum membrane. This reaction is crucial for the release of glucose into the bloodstream, and its regulation is a key point in maintaining blood glucose homeostasis (Baynes and Dominiczak, 2009).

The regulation of gluconeogenesis is complex and involves hormonal control, primarily from insulin and glucagon, which have opposing effects on the pathway. Insulin, which is released in response to high blood glucose levels, inhibits gluconeogenesis by decreasing the expression of key enzymes such as PEPCK and glucose-6-phosphatase, whereas glucagon promotes gluconeogenesis by increasing the transcription of these enzymes. Furthermore, allosteric regulation by metabolites such as acetyl-CoA, AMP, and fructose-2,6-bisphosphate ensures that gluconeogenesis is coordinated with the cell's energy state (Melkonian et al., 2024).

Gluconeogenesis is an energy-intensive process that requires six ATP (or GTP) molecules per glucose molecule synthesized, as opposed to the two ATP molecules produced by glycolysis. However, this energy expenditure is required for glucose synthesis, which is critical for survival during fasting or strenuous activity (Chourpiliadis and Mohiuddin, 2024; Melkonian et al., 2024). The interplay between glycolysis and gluconeogenesis exemplifies cells' metabolic flexibility in responding to changing energy demands and nutrient availability.

3.3. CITRIC ACID CYCLE AND OXIDATIVE PHOSPHORYLATION

The citric acid cycle, also known as the Krebs cycle or the tricarboxylic acid (TCA) cycle, and oxidative phosphorylation are two interconnected processes that are essential to cellular respiration. These pathways are essential for energy production, allowing cells to convert energy stored in carbohydrates, fats, and proteins into adenosine triphosphate (ATP), the cell's universal energy currency. Both processes occur within the mitochondria, demonstrating the organelle's importance in cellular metabolism (Akram, 2014). The citric acid cycle generates high-energy electron carriers, which are used by oxidative phosphorylation to drive ATP synthesis via the electron transport chain and chemiosmosis. Understanding these pathways necessitates a thorough examination of the enzymatic reactions and regulatory mechanisms that govern them, as they are critical for maintaining cellular energy homeostasis.

The citric acid cycle begins with the entry of acetyl-CoA, a two-carbon molecule produced by the oxidative decarboxylation of pyruvate, into the mitochondria. Acetyl-CoA combines with a four-carbon molecule, oxaloacetate, to form citrate, a six-carbon tricarboxylic acid. Citrate synthase, a highly regulated enzyme, catalyzes this reaction, ensuring that the cycle only proceeds when there is enough acetyl-CoA available, reflecting the cell's energy status. The formation of citrate is the first step in the cycle, which will gradually oxidize the carbon atoms of the acetyl group to carbon dioxide while capturing high-energy electrons in the form of NADH and FADH2 (Alberts et al., 2015). Aconitase, an enzyme that facilitates the rearrangement of atoms within the molecule, then converts citrate to isocitrate, preparing it for subsequent oxidation. Isocitrate is oxidatively decarboxylated by isocitrate dehydrogenase, producing α-ketoglutarate, a five-carbon molecule, along with the first molecule of NADH and the release of carbon dioxide. This reaction is a critical regulatory step in the citric acid cycle, with isocitrate dehydrogenase being allosterically activated by ADP and inhibited by ATP and NADH, ensuring that the cycle's activity is modulated in response to the cell's energy requirements (Angumeenal and Venkappayya, 2013).

The next step involves the conversion of α-ketoglutarate to succinyl-CoA by the α-ketoglutarate dehydrogenase complex, another critical regulatory point in the cycle. This reaction is analogous to the pyruvate dehydrogenase complex and involves the decarboxylation of α-ketoglutarate, coupled with the reduction of NAD+ to NADH and the release of a second molecule of carbon dioxide. The resulting succinyl-CoA, a four-carbon molecule, is then converted to succinate by succinyl-CoA synthetase in a reaction that generates ATP or GTP via substrate-level phosphorylation (Baynes and Dominiczak, 2009). This direct synthesis of a high-energy phosphate compound is significant because it provides a link between the TCA cycle and the cell's immediate energy needs.

Succinate, now a four-carbon molecule, is oxidized to fumarate by succinate dehydrogenase, an enzyme found only in the inner mitochondrial membrane and a component of the electron transport chain. During this reaction, FAD is reduced to FADH2, which, unlike NADH, directly feeds electrons into the ETC, bypassing the first complex and adding fewer protons to the electrochemical gradient. Fumarate is then converted to malate by fumarase, which catalyzes the hydration reaction that prepares malate for the final oxidative step (Akram, 2014; Alberts et al., 2015).

Malate dehydrogenase oxidizes malate to oxaloacetate, resulting in the third and final molecule of NADH in the cycle. The regeneration of oxaloacetate is critical because it allows the citric acid cycle to continue, which allows for continuous acetyl-CoA processing. The NADH and FADH2 molecules produced during the citric acid cycle contain a lot of high-energy electrons, which are then transferred to the electron transport chain, where oxidative phosphorylation occurs (Akram, 2014).

Oxidative phosphorylation is the process by which energy stored in NADH and FADH2 is converted into ATP. This process occurs in the inner mitochondrial membrane and includes two major components: the electron transport chain and ATP synthase. The electron transport chain consists of four multiprotein complexes (Complexes I–IV) and two mobile electron carriers, ubiquinone (coenzyme Q) and cytochrome c, which shuttle electrons between the complexes (Angumeenal and Venkappayya, 2013).

The process starts with the transfer of electrons from NADH to Complex I (NADH: ubiquinone oxidoreductase), a large enzyme complex that catalyzes the transfer of two electrons from NADH to ubiquinone. During this process, four protons are pumped from the mitochondrial matrix into the intermembrane space, adding to the proton gradient that drives ATP synthesis. Ubiquinone, now reduced to ubiquinol, transfers electrons to Complex III (cytochrome BC1), which also pumps protons across the membrane and transports electrons to cytochrome C (Blumwald et al., 2000).

Complex II (succinate: ubiquinone oxidoreductase) aids in oxidative phosphorylation by oxidizing FADH2, which is produced in the citric acid cycle during the oxidation of succinate to fumarate. Unlike Complex I, Complex II does not contribute to the proton gradient and instead transfers electrons directly to ubiquinone, which transports them to Complex III (Angumeenal and Venkappayya, 2013). This establishes a direct link between the TCA cycle and the electron transport chain, emphasizing the integrative nature of cellular respiration.

Cytochrome c, a small heme-containing protein, moves electrons from Complex III to Complex IV (cytochrome c oxidase), where the final step of electron transfer takes place. Complex IV catalyzes the transfer of electrons to molecular oxygen, the terminal electron acceptor, converting it into water. This step is critical because the reduction of oxygen is accompanied by the pumping of additional protons into the intermembrane space, thereby increasing the proton gradient (Bonner and Varner, 2012).

The energy stored in the electrochemical proton gradient, also known as the proton-motive force, is used by ATP synthase (Complex V) to generate ATP from ADP and inorganic phosphate. ATP synthase is a remarkable molecular machine that works through chemiosmosis, a process in which the flow of protons back into the mitochondrial matrix via the enzyme's proton channel drives ADP phosphorylation. This process is extremely efficient, with one molecule of NADH resulting in the production of approximately 2.5 molecules of ATP and one molecule of FADH2 yielding approximately 1.5 molecules of ATP (Angumeenal and Venkappayya, 2013).

The regulation of oxidative phosphorylation is closely related to the cell's energy requirements, with ATP and ADP serving as key regulators. When there is a high demand for ATP, the proton gradient dissipates more quickly, which accelerates the flow of electrons through the electron transport chain and increases oxygen consumption. When ATP levels are sufficient, oxidative phosphorylation slows, allowing ATP production and consumption to be balanced (Akram, 2014). Furthermore, the process is influenced by a number of other factors, including the availability of substrates such as NADH and FADH2, the integrity of the mitochondrial membrane, and the presence of uncoupling proteins capable of dissipating the proton gradient without producing ATP, a mechanism involved in thermogenesis.

The citric acid cycle and oxidative phosphorylation are not only essential for energy production but also play important roles in other cellular processes. Intermediates of the TCA cycle act as precursors for the biosynthesis of amino acids, nucleotides, and other macromolecules, connecting energy metabolism to anabolic pathways. The regulation of these pathways is critical for cellular function, and disruptions can result in a variety of metabolic disorders, including mitochondrial diseases, which are frequently characterized by impaired oxidative phosphorylation (Albert et al., 2015).

3.4. PHOTOSYNTHESIS IN PLANTS

Photosynthesis is a basic biochemical process in which green plants, algae, and some bacteria convert light energy into chemical energy stored in glucose and other organic molecules. This process is critical for life on Earth because it provides the primary source of organic matter for almost all organisms, either directly or indirectly. Photosynthesis occurs in the chloroplasts of plant cells, where light energy is used to drive the synthesis of organic compounds from carbon dioxide and water, with oxygen released as a byproduct. Photosynthesis consists of two major stages: light-dependent reactions and light-independent reactions, also known as the Calvin cycle (Pessarakli, 2024). These stages are intricately linked and together enable the conversion of solar energy into a stable chemical form (Figure 3.2).

The light-dependent reactions, also known as photochemical reactions, occur in the chloroplasts' thylakoid membranes. These reactions start with the absorption of light by chlorophyll, a pigment that captures light energy most effectively in the blue and red regions of the electromagnetic spectrum. Chlorophyll molecules are organized into photosystems, which are classified into two types: photosystem II (PSII) and photosystem I (PSI). Each photosystem is a light-harvesting complex that absorbs photons and transfers the energy to a specific pair of chlorophyll a molecules at the reaction center. The absorption of light energy in Photosystem II causes an electron in the reaction center chlorophyll to become more energetic (Hohmann-Marriott and Blankenship, 2011). This high-energy electron is then transferred to a series of electron carriers in the electron transport chain.

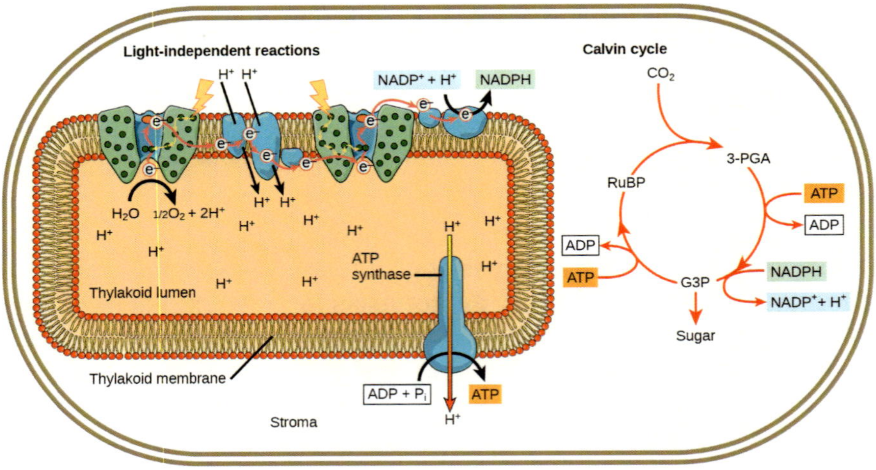

Figure 3.2. *The two major stages of photosynthesis.*

Source https://courses.lumenlearning.com/suny-wmopen-biology1/chapter/photosynthesis/.

The initial electron transfer from PSII is accompanied by the photolysis of water molecules, which is the process of separating water into oxygen, protons, and electrons. The oxygen-evolving complex associated with PSII catalyzes this reaction, supplying the electrons needed to replace those lost by chlorophyll in the reaction center. Water splitting generates oxygen as a byproduct, which diffuses from the chloroplast and eventually enters the atmosphere. The produced protons contribute to the proton gradient across the thylakoid membrane, which is necessary for ATP synthesis (Bonner and Varner, 2012).

As the excited electron from PSII travels down the electron transport chain, it loses energy, which is then used to pump protons from the stroma into the thylakoid lumen. This generates a proton gradient, or proton-motive force, across the thylakoid membrane. The energy stored in this gradient is then used by ATP synthase, an enzyme that produces ATP from ADP and inorganic phosphate through a process known as photophosphorylation. The chloroplasts produce ATP, the cell's energy currency, which is then used in light-independent reactions (Bonner and Varner, 2012).

The electron, now at a lower energy level, is transferred to Photosystem I, where it is once again excited by light absorbed by the PSI reaction center. The re-energized electron travels through a second electron transport chain before being transferred to the electron carrier NADP+ to form NADPH. NADPH, together with ATP, provides the reducing power and energy required for glucose synthesis in the Calvin cycle. The light-dependent reactions produce the energy-rich molecules ATP and NADPH, which are required for the next stage of photosynthesis when carbon dioxide is converted into organic compounds (Hohmann-Marriott and Blankenship, 2011; Pessarakli, 2024).

The light-independent reactions, also known as the Calvin cycle, occur in the chloroplast stroma and do not require light directly; however, they rely on the ATP and NADPH produced by the light-dependent reactions. The Calvin cycle is a series of enzymatic reactions that convert carbon dioxide into organic molecules, eventually yielding glucose. The cycle has three major phases: carbon fixation, reduction, and regeneration of the starting molecule, ribulose-1,5-bisphosphate (RuBP) (Hohmann-Marriott and Blankenship, 2011).

The first phase of the Calvin cycle, carbon fixation, starts with the incorporation of carbon dioxide into an organic molecule. The enzyme ribulose-1,5-bisphosphate carboxylase/oxygenase, also known as Rubisco, is the most abundant enzyme on Earth and catalyzes this reaction. Rubisco catalyzes the carboxylation of RuBP, a five-carbon sugar, by adding carbon dioxide, resulting in a transient six-carbon intermediate. This intermediate quickly breaks down into two molecules of 3-phosphoglycerate (3-PGA), a three-carbon compound. Carbon dioxide fixation is an important step in the Calvin cycle because it converts inorganic carbon to an organic form that the cell can then process (Pessarakli, 2024).

In the second phase, reduction, the 3-PGA molecules are phosphorylated by ATP and then reduced by NADPH to form glyceraldehyde-3-phosphate (G3P). This step is analogous to similar reactions in glycolysis but operates in reverse, using the energy and reducing power supplied by the light-dependent reactions. For every three molecules of carbon dioxide that enter the cycle, six molecules of G3P are produced. However, only one G3P molecule is exported from the cycle for use in the synthesis of glucose and other carbohydrates, while the remaining five molecules are used to regenerate RuBP (Hohmann-Marriott and Blankenship, 2011).

The Calvin cycle's third phase, regeneration, consists of a series of reactions that rearrange the carbon skeletons of the five G3P molecules to regenerate three molecules of RuBP. This regeneration process requires more ATP and is necessary for the cycle to continue. The regenerated RuBP is then able to accept new carbon dioxide molecules, allowing the cycle to continue. As a result, as long as there is a supply of ATP, NADPH, and carbon dioxide, the Calvin cycle continues to function, producing glucose and other carbohydrates that the plant can use for energy and growth (Bonner and Varner, 2012).

Photosynthesis is a highly regulated process with many variables influencing its efficiency and rate. Environmental conditions, such as light intensity, temperature, and the availability of water and nutrients, all have a significant impact on a plant's overall photosynthesis activity. Furthermore, the regulation of key enzymes, such as Rubisco, and the coordination of light-dependent and light-independent reactions ensure that photosynthesis occurs efficiently and that the plant's energy requirements are met (Pessarakli, 2024).

Photosynthesis not only sustains plants, but it also supports life on Earth by producing oxygen and organic matter. Photosynthesis converts solar energy into a form that can be used by heterotrophic organisms, including humans, and thus serves as the foundation of most food webs. Furthermore, photosynthesis is critical to the global carbon cycle because it helps to sequester carbon dioxide from the atmosphere, reducing the impact of greenhouse gases on climate change (Hohmann-Marriott and Blankenship, 2011; Pessarakli, 2024).

3.5. LIPID METABOLISM

Lipid metabolism is a crucial mechanism in cell biochemistry that controls cellular energy, membrane structure, signaling, and homeostasis. Lipids, which include fatty acids, triglycerides, phospholipids, and sterols, are critical components of biological membranes that act as energy reserves (Smith et al., 2018). The metabolic pathways that govern lipid production, storage, and degradation are tightly controlled to preserve cellular function and organismal health. Abnormalities in lipid metabolism are related to several metabolic disorders, including obesity, cardiovascular diseases, and diabetes, emphasizing the necessity of understanding the molecular pathways involved (Chandel, 2021).

The synthesis and breakdown of fatty acids and their derivatives are critical in cellular lipid metabolism for energy homeostasis. Lipogenesis, or fatty acid production, occurs largely in the cytoplasm of cells, particularly in the liver and adipose tissue. Acetyl-CoA, produced from carbohydrate metabolism, is the precursor to fatty acid synthesis (Baynes and Dominiczak, 2009). Acetyl-CoA is converted to malonyl-CoA by the enzyme acetyl-CoA carboxylase, which is an important regulatory step in lipogenesis (Jackson and Wang, 2018). Fatty acid synthase, a large enzyme complex, then catalyzes the extension of the carbon chain by adding two-carbon units to produce long-chain fatty acids like palmitate (Baynes and Dominiczak, 2009).

Once synthesized, fatty acids are either incorporated into triglycerides for storage in adipose tissue or used to form complex lipids such as phospholipids and sphingolipids, which are critical components of cellular membranes (Martinez et al., 2016). Triglyceride synthesis involves the esterification of three fatty acids to a glycerol backbone and occurs primarily in the endoplasmic reticulum (Pollard et al., 2024). These triglycerides are then packaged into lipoproteins, which are secreted into the bloodstream for transport to various tissues where they are either stored or oxidized for energy.

The catabolic route of lipid metabolism, known as lipolysis, breaks down triglycerides into free fatty acids and glycerol. Lipolysis is generally induced by hormonal signals, such as catecholamines, which activate lipases in adipose tissue. Once liberated, free fatty acids are transferred to mitochondria for β-oxidation. This process comprises the successive removal of two-carbon units from the fatty acid chain, resulting in acetyl-CoA, which enters the citric acid cycle and generates ATP. The energy production from fatty acid oxidation is important, particularly during periods of fasting or prolonged physical exercise, when carbohydrate stores are depleted, and the body relies on fat as its primary energy source (Baynes and Dominiczak, 2009).

Cholesterol metabolism is another crucial aspect of lipid biochemistry. Cholesterol, a sterol molecule, is synthesized in the liver and is an essential component of cell membranes, providing structural integrity and fluidity. It also serves as a precursor for the synthesis of steroid hormones, bile acids, and vitamin D. Cholesterol biosynthesis begins with acetyl-CoA and proceeds through a series of enzymatic reactions, with 3-hydroxy-3-methylglutaryl-CoA reductase (HMG-CoA reductase) being the rate-limiting enzyme (Smith et al., 2018). This enzyme is the target of statins, a class of drugs used to lower cholesterol levels in individuals with hypercholesterolemia (Feng et al., 2016). Once synthesized, cholesterol can be esterified and stored within cells or transported via lipoproteins to peripheral tissues (Smith et al., 2018).

Hormonal and dietary cues carefully control the metabolism of fats. For example, insulin increases the expression of lipogenic enzymes, such as fatty acid synthase and acetyl-CoA carboxylase, which in turn enhances lipogenesis. On the other hand, glucagon and adrenaline facilitate the release of fatty acids from adipose tissue by activating hormone-sensitive lipase, which in turn promotes lipolysis (Liu et al., 2020). Energy storage and expenditure are carefully regulated in response to the body's metabolic demands thanks to the balance between two opposing pathways.

A significant area of research in lipid metabolism concerns the interplay between lipid signaling and metabolic diseases. Dysregulation of lipid metabolism can lead to the accumulation of lipid intermediates, such as ceramides and diacylglycerols, which are known to interfere with insulin signaling pathways, leading to insulin resistance—a hallmark of type 2 diabetes. Similarly, elevated levels of circulating free fatty acids have been implicated in the development of non-alcoholic fatty liver disease (NAFLD) and cardiovascular diseases. Understanding the molecular mechanisms underlying these processes is critical for developing therapeutic strategies to combat metabolic disorders (Baynes and Dominiczak, 2009).

Lipid metabolism is also linked to autophagy, a cellular breakdown system that breaks down and recycles cellular components in response to stress or nutritional restriction. Lipophagy, a type of autophagy that targets lipid droplets, is a growing field of research in cell biology. Lipophagy allows cells to mobilize stored lipids and provide substrates for mitochondrial oxidation, sustaining energy generation during metabolic stress. Lipophagy dysfunction has been connected with metabolic disorders, emphasizing the relevance of this system in maintaining cellular lipid homeostasis (Pollard et al., 2024).

A complex network of transcription factors, enzymes, and signaling molecules regulates lipid metabolism (Smith et al., 2020). Nuclear receptors known as peroxisome proliferator-activated receptors (PPARs) are essential for controlling lipid metabolism. PPARs regulate the expression of genes related to inflammation, lipogenesis, and fatty acid oxidation. By increasing fatty acid oxidation and boosting insulin sensitivity, PPAR agonists like fibrates and thiazolidinediones are used clinically to treat hyperlipidemia and type 2 diabetes (Smith et al., 2018).

3.6. AMINO ACID METABOLISM

Amino acid metabolism plays a fundamental role in cell biochemistry, representing a central aspect of cellular physiology. Amino acids, the building blocks of proteins, serve as precursors for numerous biologically essential molecules. These small organic compounds, containing both amine and carboxyl functional groups, are necessary for the synthesis of proteins, enzymes, neurotransmitters, and other critical cellular components. The metabolic pathways involving amino acids are intricate and highly regulated, ensuring that cells maintain homeostasis while responding to changing environmental conditions. The process begins with the intake of amino acids either through dietary sources or through endogenous mechanisms such as protein degradation.

Amino acids enter the cell and are directed into different metabolic processes such as the urea cycle, gluconeogenesis, and protein synthesis. Transamination is one of the first steps in amino acid metabolism, where an amino group from one amino acid is transferred to an α-keto acid to create a new amino acid. Aminotransferases, as the reaction's catalysts, are essential to the metabolism of every amino acid. It makes it possible for nitrogen to be directed toward the production of urea through the deamination of amino acids, which mostly takes place in the liver. harmful nitrogenous waste cannot build up in the body because the urea cycle transforms harmful ammonia into urea, which the kidneys can safely eliminate (Baynes and Dominiczak, 2009).

The degradation of amino acids involves numerous interconnected processes, depending on whether the amino acid is glucogenic, ketogenic, or both. Glucogenic amino acids are those whose breakdown products can be utilized to produce glucose via gluconeogenesis, a process critical for keeping blood glucose levels stable during fasting. In contrast, ketogenic amino acids are broken down into intermediates such as acetyl-CoA and acetoacetate, which can be used to synthesize ketone bodies or fatty acids. Leucine and lysine are purely ketogenic, but isoleucine, phenylalanine, and tryptophan are both glucogenic and ketogenic (Punekar, 2018).

In addition to energy production and protein synthesis, amino acid metabolism plays an important role in immune response regulation, redox balance, and cell signaling. Glutamine and arginine, for example, are important in immune function, providing fuel to rapidly reproducing cells such as lymphocytes and macrophages. Glutamine is very crucial in maintaining redox balance because it acts as a precursor for glutathione, a key antioxidant in cells (Baynes and Dominiczak, 2009). Meanwhile, arginine is a precursor to nitric oxide, a signaling molecule that has roles in vasodilation, immunological response, and neurotransmission.

The production of non-essential amino acids within the cell is another important element of amino acid metabolism. Non-essential amino acids can be produced in the body, whereas essential amino acids must be received through diet. The manufacture of these amino acids comprises many processes that rely on the availability of metabolic intermediates such pyruvate, oxaloacetate, and α-ketoglutarate. Alanine is created from pyruvate by a transamination mechanism, while aspartate and glutamate are made from oxaloacetate and α-ketoglutarate, respectively (Punekar, 2018).

Amino acid metabolism is also tightly linked to cellular signaling networks, particularly the mTOR pathway. mTOR is a serine/threonine kinase that controls cell growth, proliferation, and survival by sensing food availability, which includes amino acids. When amino acids, particularly leucine, are abundant, mTOR activity increases, which promotes anabolic activities such as protein synthesis. In contrast, when amino acid levels are low, mTOR activity is blocked, resulting in a drop in protein synthesis and an increase in autophagy, the process by which cells breakdown and recycle their own components to maintain cellular homeostasis under food deprivation (Pollard et al., 2024).

Amino acid metabolism disruptions are linked to a variety of metabolic illnesses and diseases. For example, phenylketonuria (PKU) is a hereditary condition characterized by a lack of phenylalanine hydroxylase, the enzyme responsible for converting phenylalanine to tyrosine. This causes a buildup of phenylalanine in the blood, which can cause severe brain damage if not treated. Dietary restriction of phenylalanine is used to treat PKU and prevent harmful buildup. Similarly, illnesses such as maple syrup urine disease (MSUD) are caused by abnormalities in the metabolism of branched-chain amino acids (BCAAs), including leucine, isoleucine, and valine, leading to toxic accumulation and neurological problems (Baker, 2009).

The importance of amino acids in cell signaling goes beyond the mTOR pathway. Amino acids also function as neurotransmitters and precursors to neurotransmitter production. Glutamate, for example, is the most abundant excitatory neurotransmitter in the central nervous system, with important roles in synaptic plasticity, learning, and memory. In contrast, glycine and gamma-aminobutyric acid (GABA) act as inhibitory neurotransmitters, lowering neural excitability. Thus, amino acid metabolism is inextricably related to nervous system function, and disruptions in these pathways can result in neurological disorders such as epilepsy, schizophrenia, and neurodegenerative diseases (Baynes and Dominiczak, 2009).

Amino acid metabolism is also relevant in the context of cancer biology. Cancer cells exhibit altered metabolism, often referred to as the Warburg effect, where they preferentially use glycolysis for energy production, even in the presence of oxygen. However, amino acids play a crucial role in supporting the rapid growth and proliferation of cancer cells. Glutamine, for example, is a major fuel source for many cancer cells, providing nitrogen for nucleotide and amino acid synthesis and replenishing tricarboxylic acid (TCA) cycle intermediates through anaplerosis. Targeting glutamine metabolism has emerged as a potential therapeutic strategy in cancer treatment (Punekar, 2018).

The metabolism of sulfur-containing amino acids, such as methionine and cysteine, is also critical in cellular function. Methionine is a precursor to S-adenosylmethionine (SAM), which is an important methyl donor in methylation events that control gene expression, protein function, and lipid metabolism. Cysteine, on the other hand, is involved in the synthesis of glutathione and taurine, two molecules required for antioxidant defense and bile salt generation, respectively. Sulfur amino acid metabolism dysregulation has been linked to multiple disorders, including cardiovascular disease and dementia (Zhang et al., 2018).

3.7. REGULATION OF METABOLIC PATHWAYS

Maintaining cellular homeostasis and adjusting to changing environmental conditions depend heavily on the control of metabolic pathways in cell biochemistry. Metabolic pathways refer to intricate networks of chemical reactions that facilitate biological functions such as energy production, biosynthesis, and degradation by converting substrates into products via a sequence of enzyme steps. For the cell to effectively manage energy resources, adapt to changes in food availability, and maintain optimal cellular function, these pathways are regulated at numerous levels. A highly dynamic and flexible metabolic landscape is produced by the intricate interactions between enzymes, metabolites, signaling molecules, and transcriptional factors that govern these pathways.

One of the primary modes of metabolic regulation occurs through allosteric control of enzymes, where enzymes are modulated by small molecules that bind to sites distinct from the active site, thus altering their activity. A classic example of this form of regulation is found in glycolysis, where the enzyme phosphofructokinase-1 (PFK-1) is allosterically inhibited by ATP and activated by AMP. This ensures that glycolysis is downregulated when cellular ATP levels are sufficient and upregulated when energy is scarce (Pollard et al., 2024). Such feedback mechanisms are essential for maintaining cellular energy balance. Similarly, many enzymes in the citric acid cycle are regulated by the levels of their substrates and products, allowing the cycle to be finely tuned according to the cell's metabolic needs.

Hormonal regulation is another important mechanism by which cells control their metabolic processes. Insulin, glucagon, and epinephrine operate as metabolic regulators, regulating critical processes such as glycolysis, gluconeogenesis, glycogen synthesis, and lipid metabolism. Insulin, for example, stimulates glucose absorption in cells and glycolysis by activating enzymes like hexokinase and pyruvate kinase (Bonner and Varner, 2012). In contrast, glucagon increases gluconeogenesis and glycogenolysis in the liver, elevating blood glucose levels during fasting or low-energy situations. These hormonal signals are mediated by cell surface receptor proteins that initiate intracellular signaling cascades, which frequently include phosphorylation and dephosphorylation events that alter the activity of target enzymes.

Phosphorylation is a key post-translational change that regulates metabolic enzymes. Protein kinases, such as AMP-activated protein kinase (AMPK), act as metabolic sensors, phosphorylating and regulating specific enzyme activity in response to variations in cellular energy status (Alberts et al., 2015). When energy levels are low, AMPK is activated and phosphorylates enzymes engaged in anabolic pathways, blocking processes like fatty acid and protein synthesis. Concurrently, AMPK activates catabolic pathways such as fatty acid oxidation and autophagy to restore energy balance. The capacity of AMPK to integrate energy status with metabolic control highlights the significance of reversible phosphorylation as a quick and effective way to regulate metabolism in response to cellular conditions.

By regulating the expression of genes producing metabolic enzymes, transcriptional regulation also contributes significantly to long-term metabolic adaptation. For example, the transcription factor peroxisome proliferator-activated receptor alpha (PPARα) is activated in the liver in response to fasting, which results in the overexpression of genes related to ketogenesis and fatty acid oxidation (Bonner and Varner, 2012). On the other hand, transcription factors such as sterol regulatory element-binding proteins (SREBPs) promote the expression of genes related to lipid production when nutrients are abundant (Punekar, 2018). Cells can adjust their metabolic potential for prolonged periods of time through transcriptional regulation, especially in response to dietary modifications, hormone cues, or environmental stresses.

Maintaining metabolic homeostasis requires not only the control of specific enzymes and pathways but also the synchronization of several metabolic processes. AMPK, mTOR (mechanistic target of rapamycin), and sirtuins are examples of metabolic "master regulators" that frequently work together to achieve this coordination. To orchestrate a general metabolic response, these regulatory proteins integrate information from energy levels, stress responses, and nutrients. For example, in response to increasing nutrition availability, the mTOR pathway inhibits autophagy and promotes protein synthesis, acting as a crucial regulator of cellular development and metabolism (Alberts et al., 2015). Conversely, under conditions of nutrient deprivation, mTOR is inhibited, leading to a shift towards catabolic processes such as autophagy, which helps the cell survive by degrading and recycling cellular components.

Furthermore, metabolic pathways do not act in isolation but are linked by common intermediates and co-regulation. The balance between glycolysis and oxidative phosphorylation, known as the "Warburg effect" in cancer cells, exemplifies how changes in metabolic flux might be linked to cellular proliferation and survival (Baynes and Dominiczak, 2009). Even under aerobic settings, glycolysis is enhanced in cancer cells, allowing for rapid biosynthetic precursor production despite a less effective method of ATP generation. Oncogenes and tumor suppressors cause metabolic reprogramming by regulating important metabolic enzymes, emphasizing the link between cellular signaling pathways and metabolic regulation.

Metabolite-sensing mechanisms provide another layer of regulation, allowing cells to fine-tune their metabolic pathways in real-time. Certain metabolites, such as acetyl-CoA, NADH, and AMP, serve as signaling molecules, informing the cell about its metabolic state and influencing enzyme activity accordingly. Acetyl-CoA, for example, is both a substrate for the citric acid cycle and a signal that activates acetylation-dependent transcription factors, which control genes involved in energy metabolism and lipid biosynthesis (Bonner and Varner, 2012). Metabolites' dual role emphasizes their relevance as substrates and regulators in metabolic networks.

Cellular compartmentalization, which guarantees that particular processes take place in discrete organelles like mitochondria, peroxisomes, and the cytoplasm, also has an impact on the control of metabolic pathways. This compartmentalization enables the creation of distinct microenvironments that are favorable to particular enzymatic activity in addition to physically dividing conflicting metabolic processes. For instance, to avoid pointless cycles of synthesis and breakdown, fatty acid oxidation mostly happens in the mitochondria, whereas fatty acid synthesis occurs in the cytoplasm (Baynes and Dominiczak, 2009). Furthermore, different pools of metabolites and cofactors can be housed within various cellular compartments, which improves the regulation and specificity of metabolic processes.

Interorgan communication is another type of metabolic regulation that maintains systemic energy balance. Different tissues, including the liver, muscle, adipose tissue, and brain, specialize in specific metabolic tasks and communicate with one another via hormones, metabolites, and cytokines. The liver, for example, serves as the fundamental hub of glucose metabolism, regulating blood glucose levels during fasting via glycogenolysis and gluconeogenesis and storing glucose as glycogen during eating. Adipose tissue, on the other hand, regulates lipid metabolism by releasing fatty acids when energy is low and storing triglycerides when energy is abundant. These tissues are inextricably linked by hormonal signals like insulin and leptin, which regulate their metabolic activities to meet the body's total energy requirements (Pollard et al., 2024).

Cells' responses to stress, such as oxidative stress, food deprivation, or hypoxia, demonstrate the dynamic nature of metabolic control. Under these conditions, cells change their metabolism to aid in survival and recovery. Hypoxic circumstances stabilize the hypoxia-inducible factor 1-alpha (HIF-1α), which upregulates glycolytic enzymes and suppresses oxidative phosphorylation (Alberts et al., 2015). This change to anaerobic metabolism enables cells to produce ATP in the absence of sufficient oxygen while reducing the formation of reactive oxygen species (ROS). Similarly, under oxidative stress, cells engage antioxidant defense mechanisms such as the pentose phosphate pathway, which produces NADPH for glutathione reduction and ROS neutralization (Bonner and Varner, 2012).

Chapter 4

Signal Transduction

Contents

Signal transduction is a fundamental process in cellular biochemistry that converts extracellular signals into specific intracellular responses. This intricate network of pathways ensures that cells perceive and respond appropriately to various environmental cues, maintaining homeostasis and facilitating cellular communication (Alberts et al., 2015).

The interaction of ligands (such as hormones, growth factors, and neurotransmitters) with their corresponding cell surface receptors is central to signal transduction. When ligands bind to these receptors, they undergo conformational changes, triggering a cascade of intracellular signaling events. These events frequently involve the activation of second messengers, protein kinases, and other regulatory proteins that amplify the signal, resulting in specific cellular outcomes such as gene expression, metabolic changes, or apoptosis. The specificity and efficiency of signal transduction pathways are critical because dysregulation can result in pathological conditions such as cancer and autoimmune diseases (Pollard et al., 2022) (Figure 4.1).

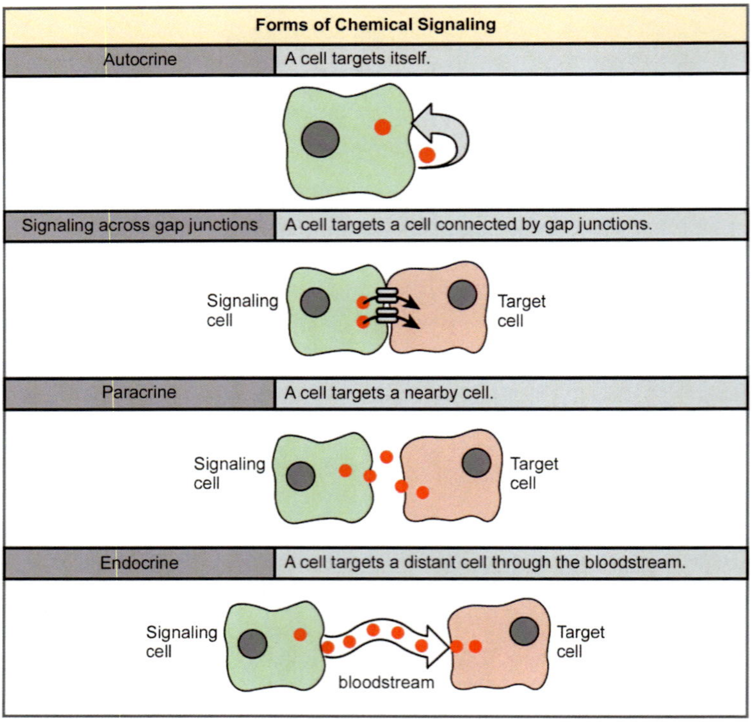

Figure 4.1. *Stages of cell signaling.*

Source: https://courses.lumenlearning.com/suny-wmopen-biology1/chapter/signaling-molecules-and-cellular-receptors/.

Advances in molecular biology have revealed key aspects of these pathways, such as the role of G protein-coupled receptors and receptor tyrosine kinases, emphasizing their significance in therapeutic targeting (Boal, 2012). Understanding the nuances of signal transduction is thus critical for developing novel interventions that modulate cellular responses in disease settings.

4.1. CELL SIGNALING MOLECULES

Cell signaling molecules are essential components of the intricate communication networks that govern cellular function, ensuring the coordination and regulation of biological processes. These molecules act as messengers, transmitting signals from one cell to another or within a cell, enabling it to respond to external stimuli, maintain homeostasis, and adapt to changing environments (Alberts et al., 2015). The diversity and specificity of these signaling molecules are critical to their role in orchestrating a wide range of cellular activities, including growth and differentiation, immune responses, and apoptosis.

Hormones, such as steroid and peptide hormones, are among the most common types of signaling molecules. Steroid hormones, such as estrogen and testosterone, are lipophilic molecules that can easily pass through the cell membrane and bind to intracellular receptors located in the cytoplasm or nucleus. This binding activates the receptor, causing changes in gene expression that trigger long-term cellular responses. In contrast, peptide hormones like insulin and glucagon are hydrophilic and cannot cross the cell membrane. They act by binding to cell surface receptors, triggering intracellular signaling cascades that frequently involve second messengers such as cyclic AMP (cAMP) or calcium ions, which further propagate the signal (Alberts et al., 2015; Boal, 2012; Chhichholiya et al., 2024).

Neurotransmitters are another important class of signaling molecules which allow neurons to communicate with other cells. In response to an action potential, neurotransmitters such as dopamine, serotonin, and acetylcholine are released from the presynaptic neuron and travel across the synaptic cleft to bind to receptors on postsynaptic cells. Depending on the neurotransmitter and receptor involved, this binding can stimulate or inhibit neuronal activity (Boal, 2012). The rapid and transient nature of neurotransmitter signaling is required for precise neuronal circuit regulation as well as the maintenance of cognitive and motor functions (Pollard et al., 2022; Chhichholiya et al., 2024).

Cytokines are another type of signaling molecules that play an important role in the immune system by mediating communication between immune cells. Cells such as macrophages, T cells, and fibroblasts produce these small proteins, which include interleukins, interferons, and tumor necrosis factors. Cytokines bind to specific receptors on target cells, activating signaling pathways that control immune responses, inflammation, and cell proliferation. Cytokine signaling balance and regulation are critical, as dysregulation can result in inflammatory diseases, autoimmunity, and immunodeficiency (Pollard et al., 2022).

Growth factors are also important signaling molecules, influencing cell proliferation, differentiation, and survival. Examples include epidermal growth factor (EGF), which binds to the epidermal growth factor receptor (EGFR) and initiates a series of intracellular signals that promote cell division and wound healing. Similarly, vascular endothelial growth factor (VEGF) is important in angiogenesis because it stimulates the formation of new blood vessels. Growth factor-activated signaling pathways frequently involve receptor tyrosine kinases, which undergo autophosphorylation upon ligand binding and recruit downstream signaling proteins that propagate the signal to the nucleus, influencing gene expression (Alberts et al., 2015; Chhichholiya et al., 2024).

In addition to these well-studied classes, lipid-based signaling molecules like eicosanoids and sphingolipids play important roles in inflammation, cell growth, and apoptosis. Eicosanoids, which include prostaglandins and leukotrienes, are derived from arachidonic acid and act locally in a paracrine or autocrine manner to mediate inflammation and immune responses. Sphingolipids, on the other hand, regulate cell growth, differentiation, and programmed cell death, with sphingosine-1-phosphate serving as a notable example that influences vascular development and immune cell trafficking (Chhichholiya et al., 2024).

Second messengers are another type of signaling molecules that relay and amplify the signals received by cell surface receptors. Cyclic nucleotides such as cAMP and cyclic GMP (cGMP) are among the most studied second messengers, influencing a wide range of physiological processes by activating protein kinases and ion channels. Calcium ions (Ca^{2+}) regulate muscle contraction, neurotransmitter release, and gene transcription by interacting with calcium-binding proteins like calmodulin (Boal, 2012; Pollard et al., 2022).

The precise interactions between these molecules and their receptors, as well as the temporal and spatial regulation of signaling pathways, enable cell signaling to be specific and coordinated. Complex feedback mechanisms, such as positive and negative feedback loops, keep cellular responses finely tuned to avoid excessive or insufficient signaling, which can lead to pathological conditions. For example, the Ras-MAPK pathway, which is activated by growth factors, is tightly regulated to prevent uncontrolled cell proliferation, which is characteristic of cancer (Alberts et al., 2015; Chhichholiya et al., 2024).

4.2. RECEPTORS AND SIGNAL TRANSDUCTION PATHWAYS

Cell surface receptors are critical components of cellular signaling networks, serving as primary mediators between extracellular signals and intracellular responses. These receptors are located in the plasma membrane and detect a wide range of external stimuli, including hormones, growth factors, and neurotransmitters. Upon ligand binding, these receptors undergo conformational changes that initiate intracellular signaling cascades, resulting in specific cellular outcomes. G protein-coupled receptors (GPCRs), receptor tyrosine kinases (RTKs), ion channel receptors, and cytokine receptors are some of the most common cell surface receptors (Boal, 2012). Each class has a unique role in cellular signaling, and their signal transduction pathways are essential for maintaining cellular function and homeostasis.

G protein-coupled receptors (GPCRs) are the most numerous and diverse family of cell surface receptors. They are distinguished by their seven-transmembrane domain structure and ability to activate heterotrimeric G proteins upon ligand binding. GPCRs respond to a wide variety of ligands, including hormones, neurotransmitters, and sensory stimuli. When activated, the receptor changes its conformation to interact with the associated G protein and activate it by promoting the exchange of GDP for GTP on the Gα subunit (Alberts et al., 2015)

The activated Gα subunit and dissociated Gβγ subunits can interact with downstream effectors like adenylate cyclase, phospholipase C, or ion channels, resulting in the production of second messengers like cAMP or inositol trisphosphate (IP3). These second messengers propagate the signal by activating protein kinases such as protein kinase A (PKA) and protein kinase C (PKC), which eventually cause changes in cellular functions like gene expression, metabolism, and cytoskeletal reorganization. The versatility and widespread presence of GPCRs make them important drug development targets, with a significant proportion of pharmaceuticals designed to modulate GPCR activity (Alberts et al., 2015; Chhichholiya et al., 2024).

Receptor tyrosine kinases (RTKs) are another important type of cell surface receptor that regulate cell growth, differentiation, and survival. RTKs are single-pass transmembrane proteins that have both an extracellular ligand-binding domain and an intracellular tyrosine kinase domain. Ligand binding, which typically involves growth factors such as epidermal growth factor (EGF) or platelet-derived growth factor (PDGF), stimulates receptor dimerization, bringing the intracellular kinase domains closer, and facilitating autophosphorylation on specific tyrosine residues. These phosphorylated tyrosines serve as docking sites for various signaling proteins containing SH2 (Src homology 2) or PTB (phosphotyrosine-binding) domains (Pollard et al., 2022).

The recruited proteins, including adaptor proteins like Grb2 and enzymes like phosphoinositide 3-kinase (PI3K) or phospholipase Cγ (PLCγ), initiate downstream signaling cascades such as the Ras-MAPK (mitogen-activated protein kinase) and the PI3K-Akt pathways. The Ras-MAPK pathway is essential for cell proliferation and differentiation, with Ras activation initiating a kinase cascade that results in the phosphorylation and activation of MAPKs such as ERK1/2, which then translocate to the nucleus and regulate gene expression. The PI3K-Akt pathway, on the other hand, stimulates cell survival and growth by activating Akt, which phosphorylates a number of substrates involved in apoptosis, metabolism, and protein synthesis. Many cancers exhibit dysregulation of RTK signaling, making these receptors and their pathways key targets for anti-cancer therapies (Gomperts et al., 2009; Pollard et al., 2022).

Ion channel receptors are another important type of cell surface receptor that cause rapid and transient changes in cell membrane potential in response to specific ligands. These receptors, also known as ligand-gated ion channels, are most commonly found in neurons and muscle cells, where they play important roles in neurotransmission and muscle contraction. Examples are the nicotinic acetylcholine receptor (nAChR) and the N-methyl-D-aspartate receptor (NMDAR). When ligands like acetylcholine or glutamate bind to these receptors, they undergo conformational changes that open an ion channel. This allows specific ions like Na^+, K^+, Ca^{2+}, or Cl^- to flow across the membrane. The resulting membrane potential changes can initiate action potentials in neurons or trigger muscle contractions in muscle cells. The nervous and muscular systems require the rapid response and specificity of ion channel receptors to function properly (Gomperts et al., 2009; Alberts et al., 2015).

Cytokine receptors are a type of cell surface receptors that primarily regulate immune responses by binding to cytokines like interleukins, interferons, and tumor necrosis factors. These receptors typically communicate via the Janus kinase (JAK)-signal transducer and activator of transcription (STAT) pathway. Upon cytokine binding, the receptor dimerizes, bringing associated JAKs into close proximity, which leads to their activation through transphosphorylation. Activated JAKs then phosphorylate specific tyrosine residues on the receptor, forming docking sites for STAT proteins. Once recruited to the receptor, JAKs phosphorylate STATs, leading to their dimerization and translocation to the nucleus, where they function as transcription factors to regulate the expression of genes involved in immune responses, cell growth, and apoptosis. The JAK-STAT pathway is tightly regulated, with negative feedback mechanisms such as the expression of suppressor of cytokine signaling (SOCS) proteins, which inhibit JAK activity or STAT docking (Chhichholiya et al., 2024).

In addition to these major classes, several other cell surface receptors and signaling pathways add to the complexity and specificity of cellular communication. Examples of receptors that activate distinct signaling pathways with critical biological functions include Notch receptors, which are involved in cell fate determination, and Toll-like receptors (TLRs), which play an important role in innate immunity. Cells can respond to a wide range of external cues with remarkable precision and adaptability due to the integration and cross-talk of different signaling pathways (Boal, 2012; Alberts et al., 2015).

4.3. PROTEIN KINASES AND PHOSPHATASES

Protein kinases and phosphatases are key regulators of cellular processes, including signal transduction, cell cycle control, metabolism, and apoptosis. Their coordinated actions ensure that cellular responses to external stimuli are properly modulated, allowing for precise control over different biological pathways. Protein kinases accomplish this regulation by catalyzing the transfer of a phosphate group from ATP to specific amino acid residues, usually serine, threonine, or tyrosine, on target proteins. This post-translational modification frequently causes conformational changes in the substrate protein, affecting its activity, localization, or interaction partners and thus modulating its function. In contrast, protein phosphatases remove these phosphate groups, reversing the effects of kinases and ensuring that phosphorylation signals are transient and tightly regulated (Gomperts et al., 2009).

Protein kinases play an important role in signal transduction as they participate in a variety of signaling pathways that regulate key cellular processes. The mitogen-activated protein kinase (MAPK) family is one of the most extensively studied kinase families, as it regulates cell proliferation, differentiation, and survival. Growth factors, cytokines, and other extracellular signals that bind to cell surface receptors, such as receptor tyrosine kinases, commonly activate the MAPK signaling pathway. Upon activation, these receptors stimulate a kinase cascade involving the sequential activation of three kinases: a MAP kinase kinase kinase (MAPKKK), a MAP kinase kinase (MAPKK), and finally, a MAP kinase (MAPK). The MAPKs, which include *ERK1/2, JNK*, and *p38*, then move to the nucleus and phosphorylate various transcription factors, causing changes in gene expression that drive cellular responses. The specificity of MAPK signaling is determined by the specific combination of MAPKKKs, MAPKKs, and MAPKs involved, as well as the spatial and temporal regulation of pathway components (Boutros et al., 2008).

Another important kinase family is the protein kinase A (PKA) family, which is activated by cyclic AMP (cAMP), a second messenger produced in response to various extracellular signals, including hormones such as adrenaline. PKA has two regulatory and two catalytic subunits; when cAMP binds to the regulatory subunits, the catalytic subunits are released and activated. Activated PKA phosphorylates a wide range of substrates, including enzymes, ion channels, and transcription factors, resulting in a variety of cellular outcomes such as increased glucose metabolism, regulation of ion fluxes, and gene transcription changes. PKA's signaling versatility is due in part to the presence of various PKA isoforms and the compartmentalization of PKA activity within the cell, which allows for precise control over its substrates and downstream effects (Boutros et al., 2008; Alberts et al., 2015).

Protein kinases are also essential for cell cycle regulation. Cyclin-dependent kinases (CDKs) are a family of serine/threonine kinases that phosphorylate key substrates involved in DNA replication, mitosis, and cell division to promote cell cycle progression. CDKs are activated by binding to specific cyclins, whose levels fluctuate throughout the cell cycle, limiting CDK activity to specific phases. CDK activity is further regulated by phosphorylation at specific sites as well as association with CDK inhibitors, which add an additional level of control over cell cycle progression. Dysregulation of CDK activity is a common feature of many cancers, highlighting the importance of these kinases in maintaining proper cell cycle control (Pollard et al., 2022).

While protein kinases are important activators of cellular processes, protein phosphatases serve as critical counterbalances, ensuring that signaling pathways are reversible and can be turned off when no longer required. Protein phosphatases are divided into three types based on their substrate specificity: serine/threonine phosphatases, tyrosine phosphatases, and dual-specificity phosphatases, which can dephosphorylate both serine/threonine and tyrosine residues. Protein phosphatase 1 (PP1) is one of the most well-studied serine/threonine phosphatases, involved in a variety of cellular functions such as glycogen metabolism, muscle contraction, and cell division. PP1 dephosphorylates key proteins in these pathways, ensuring that signals initiated by kinases are properly terminated (Boutros et al., 2008; Boal, 2012).

Protein tyrosine phosphatases (PTPs) are another important class of phosphatases that dephosphorylate tyrosine residues on target proteins. PTPs play an important role in regulating signaling pathways initiated by receptor tyrosine kinases (RTKs) and non-receptor tyrosine kinases, which are involved in cell growth, differentiation, and immune responses. For example, the PTP SHP-2 plays a role in the negative regulation of the Ras-MAPK pathway by dephosphorylating specific pathway components, thereby weakening the signal.

PTP activity dysregulation can cause aberrant signaling and has been linked to a variety of diseases, including cancer and autoimmune disorders (Chhichholiya et al., 2024).

Dual-specificity phosphatases (DUSPs) are a distinct class of phosphatases capable of dephosphorylating both serine/threonine and tyrosine residues on target proteins. These phosphatases are especially important in the regulation of MAPK signaling because they act as negative feedback regulators, dephosphorylating and inactivating MAPKs. The same signals that activate MAPKs frequently induce the expression of DUSPs, resulting in a negative feedback loop that limits the duration and intensity of MAPK signaling. This feedback regulation is critical for preventing excessive or prolonged MAPK activity, which could otherwise result in abnormal cellular responses such as uncontrolled proliferation or apoptosis (Gomperts et al., 2009; Pollard et al., 2022).

The dynamic interaction between protein kinases and phosphatases is critical for fine-tuning cellular signaling networks. The spatial and temporal localization of kinases and phosphatases, the availability of their substrates, and the presence of regulatory proteins that modulate their activity all play a role in the delicate balance of phosphorylation and dephosphorylation. Disruptions in this balance can result in a variety of pathological conditions, including cancer, diabetes, and neurodegenerative diseases, emphasizing the importance of these enzymes in maintaining cellular homeostasis (Boutros et al., 2008).

Recent advances in structural biology, proteomics, and chemical biology have shed new light on the mechanisms of kinase and phosphatase regulation, as well as their roles in disease. These advances have also aided in the development of targeted therapies that modulate kinase and phosphatase activity, opening up new avenues for the treatment of diseases associated with dysregulated signaling. For example, small molecule inhibitors of specific kinases, such as those targeting BCR-ABL in chronic myeloid leukemia, have demonstrated remarkable clinical success, highlighting the therapeutic potential of targeting these enzymes. Similarly, research is underway to develop phosphatase inhibitors and activators that can selectively modulate phosphatase activity in disease settings.

4.4. SECOND MESSENGERS IN SIGNAL TRANSDUCTION

Second messengers are small intracellular molecules that are essential for amplifying and propagating signals received by cell surface receptors during cell signaling. They act as intermediaries, translating extracellular signals like hormones, neurotransmitters, and growth factors into specific intracellular responses that allow the cell to respond quickly and appropriately to its environment. These molecules are normally produced or released in response to the activation of cell surface receptors, and their levels fluctuate rapidly, allowing them to act as transient signals that can be tightly regulated both spatially and temporally. Second messengers include cyclic nucleotides (cyclic AMP and cyclic GMP), calcium ions (Ca^{2+}), inositol trisphosphate (IP3), diacylglycerol (DAG), and nitric oxide (NO). Each plays a distinct role in various signaling pathways (Newton et al., 2016).

Cyclic nucleotides, specifically cyclic adenosine monophosphate (cAMP), are among the most extensively researched second messengers. The enzyme adenylate cyclase synthesizes cAMP from ATP and is activated by G protein-coupled receptors (GPCRs) in response to extracellular signals such as hormones and neurotransmitters. Once produced, cAMP activates protein kinase A (PKA) by binding to its regulatory subunits, allowing the active catalytic subunits to be secreted. PKA then phosphorylates several target proteins, including enzymes, ion channels, and transcription factors, altering their activity and function. The effects of cAMP are diverse and depend on the specific cellular context; for example, in liver cells, cAMP-mediated activation of PKA results in the phosphorylation and activation of enzymes involved in glycogen breakdown, thus increasing glucose availability (Gomperts et al., 2009; Alberts et al., 2015). In neurons, cAMP can regulate ion channel activity and neurotransmitter release, influencing neuronal excitability and synaptic transmission. The rapid degradation of cAMP by phosphodiesterases ensures that its signaling is tightly controlled and transient, preventing prolonged activation of downstream pathways.

Cyclic guanosine monophosphate (cGMP) is another significant cyclic nucleotide that serves as a second messenger in a variety of physiological processes. The enzyme guanylate cyclase produces cGMP following activation by nitric oxide (NO) or natriuretic peptides binding to their respective receptors. cGMP primarily acts by activating protein kinase G (PKG), which phosphorylates target proteins involved in smooth muscle relaxation, platelet aggregation, and retinal phototransduction. In the cardiovascular system, for example, cGMP-mediated PKG activation relaxes vascular smooth muscle, resulting in vasodilation and reduced blood pressure. This pathway is targeted by drugs such as nitroglycerin, which raises NO levels and thus increases cGMP production to treat conditions such as angina. Similar to cAMP, phosphodiesterases tightly regulate the levels of cGMP, which degrade cGMP to GMP, ensuring that its signaling effects are short-lived and reversible (Gomperts et al., 2009). Calcium ions (Ca^{2+}) play an important role in regulating various cellular processes, such as muscle contraction, neurotransmitter release, gene expression, and cell proliferation. The concentration of Ca^{2+} in the cytoplasm is tightly controlled by various mechanisms, including its sequestration in intracellular stores like the endoplasmic reticulum (ER) and extrusion from the cell by Ca^{+} pumps. When cell surface receptors like GPCRs or RTKs are activated, Ca^{2+} is rapidly released from the ER into the cytoplasm via channels like the inositol trisphosphate receptor (IP3R) or the ryanodine receptor. Increased cytoplasmic Ca^{2+} levels activate Ca^{2+}-binding proteins like calmodulin, which then activate downstream targets like calcium/calmodulin-dependent protein kinases (CaMKs) and phosphatases. These enzymes phosphorylate or dephosphorylate specific proteins, modulating their activity and leading to diverse cellular outcomes (Chhichholiya et al., 2024; Pollard et al., 2022).

For example, in muscle cells, the rise in Ca^{2+} levels initiates contraction by promoting the interaction between actin and myosin filaments. In neurons, Ca^{2+} entry through voltage-gated channels during an action potential triggers the release of neurotransmitters at synaptic terminals, facilitating communication between neurons. The spatial and temporal dynamics of Ca^{2+} signaling are further refined by the action of various buffering proteins and by the organization of Ca^{2+} channels into microdomains, allowing for precise control over its effects (Newton et al., 2016).

Inositol trisphosphate (IP3) and diacylglycerol (DAG) are second messengers produced by the enzyme phospholipase C (PLC), which is activated downstream of GPCRs or RTKs. Phospholipid phosphatidylinositol 4,5-bisphosphate (PIP2) is the membrane phospholipid that is cleaved by PLC. IP3 is soluble and diffuses through the cytoplasm to bind to IP3 receptors on the ER, promoting Ca^{2+} release and participating in the Ca^{2+} signaling pathway. DAG, on the other hand, remains embedded in the plasma membrane and serves as a co-activator of protein kinase C (PKC). PKC is a serine/threonine kinase that phosphorylates a diverse set of substrates involved in cell proliferation, differentiation, and apoptosis. The combined actions of IP3 and DAG thus coordinate the regulation of multiple signaling pathways, integrating signals that control critical cellular functions (Gomperts et al., 2009).

Nitric oxide (NO) is a gaseous second messenger that diffuses freely across cell membranes and regulates vascular tone, neurotransmission, and immune responses. The enzyme nitric oxide synthase (NOS) synthesizes NO from L-arginine and is activated by Ca^{2+}-calmodulin and other signaling pathways. Once produced, NO activates soluble guanylate cyclase (sGC) in target cells, causing an increase in cGMP levels and the subsequent activation of PKG. NO-mediated cGMP production in the cardiovascular system relaxes smooth muscle and dilates the blood vessels, which helps to regulate blood pressure. In the nervous system, NO acts as a neurotransmitter or neuromodulator, influencing synaptic plasticity and memory formation. NO also plays an important role in the immune system, where macrophages and other immune cells produce it to mediate antimicrobial and antitumor responses. NO's rapid diffusion and short half-life enable it to act locally and transiently, ensuring that its signaling effects are tightly confined (Gomperts et al., 2009; Newton et al., 2016).

The interaction of various second messengers allows for the integration and amplification of signals within the cell. For example, simultaneous activation of cAMP and Ca^{2+} signaling pathways can lead to synergistic effects on gene expression, as PKA and CaMKs may target the same transcription factors. Cellular responses to external stimuli can be fine-tuned through cross-talk between DAG-PKC and Ca^{+}-dependent pathways. Regulating second messenger levels and their interactions with specific downstream effectors is critical for ensuring the fidelity and specificity of cellular signaling, preventing aberrant signaling that could lead to pathological conditions such as cancer, cardiovascular disease, or neurodegeneration (Boal, 2012; Alberts et al., 2015). Ongoing research has revealed new aspects of second messenger signaling, highlighting the complexity and sophistication of these intracellular communication networks, as well as their roles in health and disease.

4.5. REGULATION OF SIGNAL TRANSDUCTION

Signal transduction is a key mechanism that allows cells to detect and respond to changes in their surroundings, maintaining tissue and organ survival and optimal function. The intricate nature of signal transduction pathways, which entail several steps such as ligand binding, intracellular signaling cascades, and the activation of target genes or cellular responses, needs a strong regulatory mechanism to maintain cellular homeostasis. Cancer, diabetes, and neurological illnesses can all be caused by improper regulation of these pathways (Jin et al., 2016).

In addition to the MAPK and PI3K/Akt pathways, many other signal transduction pathways play significant roles in a variety of physiological activities. The Wnt signaling system, for example, is critical in embryonic development, stem cell control, and tissue homeostasis (Nusse and Clevers, 2017). Dysregulation of Wnt signaling has been linked to a variety of malignancies, including colorectal cancer, in which mutations in pathway components cause uncontrolled cell growth. Similarly, the Notch signaling system is required for cell differentiation and destiny determination throughout development. Aberrant Notch signaling activity is related to T-cell acute lymphoblastic leukemia (T-ALL) and other cancers (Gomperts et al., 2009).

The regulation of these pathways frequently happens at numerous levels, ranging from modification of downstream signaling molecules to receptor-ligand contact. One important regulation mechanism is the compartmentalization of signaling components within specific subcellular regions, such as lipid rafts or endosomes. This spatial structure ensures that signaling events are carefully controlled and confined to specific cellular areas, preventing signaling pathways from being activated excessively or inappropriately (Gomperts et al., 2009). For example, upon ligand interaction, the epidermal growth factor receptor (EGFR) is endocytosed and either recycled back to the plasma membrane or destroyed in lysosomes, allowing control over the duration and strength of signaling.

A further level of control involves the alteration of receptor function by post-translational changes like sumoylation and ubiquitination. Ubiquitination, for example, designates signaling proteins or receptors for lysosomal trafficking or proteasomal destruction. In order to stop signaling and make sure that cells do not stay in a state of constant activation, this process acts as a negative regulatory mechanism. However, sumoylation can change the activity, stability, or location of transcription factors and other signaling proteins, which adds another layer of complexity to the control of signal transduction (Nusse and Clevers, 2017).

One of the most well-known and researched regulating processes in signal transduction is still phosphorylation. The action of signaling molecules is dynamically regulated by the combined action of protein kinases, which phosphorylate target proteins, and phosphatases, which dephosphorylate them. Other phosphatases, like SHP-1 and SHP-2, are essential for downregulating receptor tyrosine kinase signaling in addition to PTEN's function in the PI3K/Akt pathway (Gomperts et al., 2009). Phosphatase dysregulation can result in persistent signaling, which can aid in the onset of cancer and other illnesses.

MicroRNAs (miRNAs) are also a developing regulatory mechanism in signal transduction. These short, non-coding RNAs can control gene expression after transcription by binding to target mRNAs and inhibiting or boosting their destruction. MiRNAs have been demonstrated to affect many signal transduction pathways, including those involved in cancer, neurodegeneration, and cardiovascular disease. For example, miR-21 has been shown to target PTEN and increase tumor growth by increasing PI3K/Akt signaling in cancer cells (Gomperts et al., 2009) (Figure 4.2).

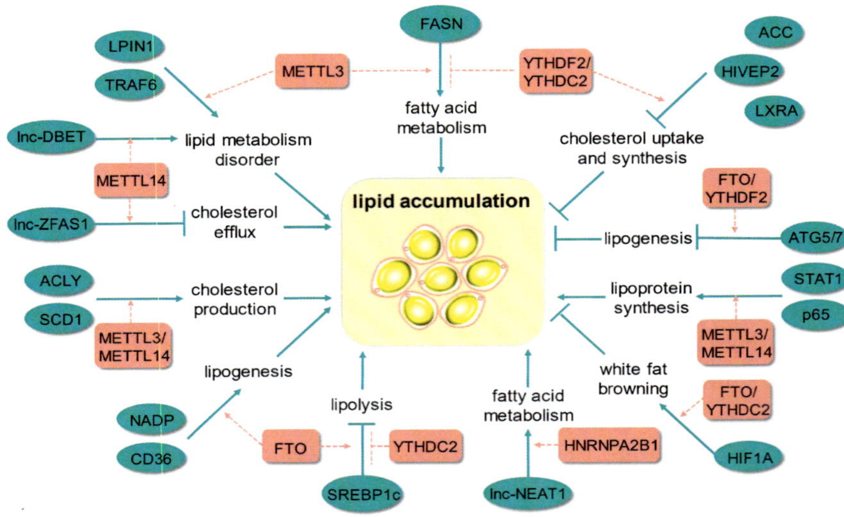

Figure 4.2. *m6A modification regulates signal transduction by destabilizing mRNA and activating ERK signaling.*

Source: Yang H, Li Y, Huang L, Fang M, Xu S. (2023). The epigenetic regulation of RNA N6-methyladenosine methylation in glycolipid metabolism. Biomolecules; 13(2):273. https://doi.org/10.3390/biom13020273.

Governing cellular responses to external stimuli through processes like desensitization and receptor downregulation is another aspect of regulating signal transmission. For instance, desensitization in G-protein-coupled receptor (GPCR) signaling happens when G-protein-coupled receptor kinases (GRKs) phosphorylate the receptor. This causes the receptor to be recruited by arrestins, which impede additional signaling and encourage receptor internalization (Nusse and Clevers, 2017). By preventing cells from becoming overstimulated by recurring signals, this procedure preserves cellular equilibrium.

Finally, an additional layer of regulation is added by the interactions between various signaling pathways. Cells can combine several impulses through crosstalk, enabling them to adjust their responses to intricate surroundings. For instance, depending on the situation, the interaction between the PI3K/Akt and MAPK pathways might affect cellular outcomes including differentiation, death, or proliferation (Jin et al., 2016). Disease progression can be facilitated by dysregulation of crosstalk processes, which can result in abnormal cellular responses.

4.6. CELL COMMUNICATION IN DEVELOPMENT

Cell communication is an important and highly choreographed process that occurs throughout development, from embryogenesis to the production and maintenance of adult tissues. It comprises a wide range of signaling channels that allow cells to communicate information, coordinate their actions, and differentiate into the specialized kinds required for organ and system function.

Cellular communication is difficult because of its various processes, which include autocrine signaling, paracrine signaling, endocrine signaling, and direct signaling across gap junctions, and the involvement of morphogens, which are diffusible chemicals that structure tissues by producing concentration gradients. Disruption of these pathways can cause developmental defects and disorders, making cell communication a key area in developmental biology (Gomperts et al., 2009) (Figure 4.3).

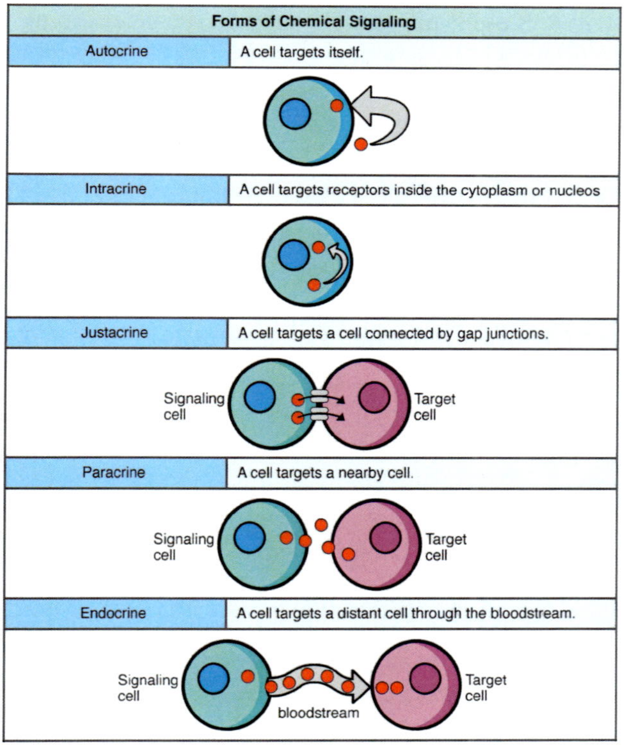

Figure 4.3. *The four categories of chemical signaling found in multicellular organisms.*

Source: By MatthewGuareschi – Own work, CC0, https://commons.wikimedia.org/w/index.php?curid=141613481.

Cell communication is responsible for ensuring that cells follow a particular developmental program that determines their fate, location, and behavior during development. This is most clear as an embryo is developing, as one fertilized egg can go through several divisions and differentiate into any one of the hundreds of different cell types that make up an organism. Growth factors, morphogens, and signaling molecules are important participants in this process because they control gene expression, tissue patterning, and cell differentiation. Transforming Growth Factor-Beta (TGF-β) and Wnt signaling pathways, among others, are essential in modulating these developmental processes (Radtke et al., 2013; Nusse and Clevers, 2017).

The Wnt signaling pathway is one of the essential elements of cell communication throughout development. Secreted ligands known as wnt proteins attach to receptors on the surface of adjacent cells to initiate intracellular signaling cascades that control the expression of certain genes. During early development, wnt signaling plays a critical role in determining cell fate and establishing the body axis (Nusse and Clevers, 2017). Wnt signaling controls the proliferation and differentiation of precursor cells in vertebrates, aiding in the development of the neural tube, somites, and limb buds. Cancer in adult tissues and developmental abnormalities including neural tube closure deficits might result from dysregulation of Wnt signaling (Jin et al., 2016).

In addition to Wnt signaling, Notch signaling is an important mechanism that regulates cell communication during development. Notch signaling is a highly conserved mechanism that controls cell fate, maintains stem cells, and promotes cell differentiation. Unlike the Wnt system, which uses diffusible ligands, Notch signaling requires direct cell-to-cell interaction. This pathway is initiated when a ligand, such as Delta or Jagged, attaches to a neighboring cell's Notch receptor. This contact causes a proteolytic cleavage of the Notch receptor, releasing the Notch intracellular domain (NICD), which translocates to the nucleus and regulates target gene expression (Jin et al., 2016). Notch signaling is critical for the development of tissues such as the neurological system, vascular system, and skin, as well as the regulation of stem cell niches. Notch signaling abnormalities have been associated with developmental problems, such as Alagille syndrome, congenital heart malformations, and some malignancies (Mollicone et al., 2009).

Morphogens are another type of chemical that facilitates cell communication during development by creating concentration gradients that offer positioning information to cells. Sonic Hedgehog (Shh) is a signaling molecule that shapes the growing neural tube and limbs, and its morphogen gradient is one of the most researched. Shh is secreted by a set of specialized cells and diffuses across the surrounding tissue, inducing varied cellular responses depending on its concentration (Briscoe and Thérond, 2013). High Shh concentrations close to the source can specify one set of cell fates, whilst lower amounts further away can specify alternative fates. This gradient process enables the precise spatial arrangement of tissues throughout development. Dysregulation of Shh signaling is linked to developmental abnormalities such as holoprosencephaly, in which the forebrain fails to divide properly, as well as tumors such as medulloblastoma (Briscoe and Thérond, 2013).

Another important phase in cell communication during development is epithelial-mesenchymal transition (EMT), a biological process in which epithelial cells lose their cell polarity and adhesion capabilities while gaining migratory and invasive qualities to become mesenchymal stem cells. EMT is essential for processes like gastrulation, neural crest formation, and heart valve development. During EMT, signaling pathways including TGF-β and Wnt influence transcription factors like Snail and Twist, resulting in downregulation of epithelial markers and upregulation of mesenchymal markers (Mollicone et al., 2009). EMT is also an important mechanism in wound healing and tissue regeneration, but when dysregulated, it can cause fibrosis and contribute to cancer metastasis.

Cell communication throughout development is also carefully controlled by feedback loops that ensure the robustness and precision of signaling events. Negative feedback mechanisms, such as phosphatases inhibiting receptor activation or inhibitors sequestering signaling molecules, contribute to the fine-tuning of signal amplitude and duration. Smad7 operates as a negative regulator in the TGF-β signaling pathway, inhibiting phosphorylation and activation of Smad2 and Smad3, dampening the signal (Gomperts et al., 2009). Positive feedback systems, on the other hand, magnify signals and are required for activities that need quick and definite cellular responses, such as the differentiation of certain cell lines.

Cells must integrate several signaling inputs during development to guarantee the appropriate creation of tissues and organs. The intercommunication among signaling pathways complicates cell connection further. For instance, the control of stem cell fate in organs like the skin and gut depends on the interaction of Wnt and Notch signaling. Wnt signaling encourages the growth of stem cells near the base of the crypt in the intestine, but Notch signaling controls the development of these cells into secretory lineages like goblet cells or absorptive enterocytes. Diseases like colorectal cancer, where aberrant Wnt signaling drives unchecked cell proliferation and defective Notch signaling modifies cell differentiation, can result from the dysregulation of this interaction (Briscoe and Thérond, 2013).

Furthermore, throughout development, mechanical stresses applied by the extracellular matrix (ECM) can affect cell communication. Through integrins and other mechanoreceptors, cells are able to detect and react to changes in the stiffness and composition of the extracellular matrix. Cell behavior, such as migration, differentiation, and proliferation, is influenced by these mechanical signals. Tissue morphogenesis and organ creation are two processes in the developing embryo where mechanical forces are involved (Gomperts et al., 2009). The regulation of cell communication during development is further complicated by the interaction between mechanical cues and biochemical signals.

4.7. DISEASES ASSOCIATED WITH SIGNALING TRANSDUCTION PATHWAYS

Signal transduction pathways are essential for cellular communication, controlling a wide range of physiological activities such as cell proliferation, apoptosis, differentiation, and metabolism. Aberrations in these pathways can cause a variety of illnesses, including cancer, neurological disorders, cardiovascular disease, and immune system malfunction. These pathways are especially vulnerable to dysregulation due to their complexity, which includes a large number of signaling molecules, receptors, and downstream effectors. This dysregulation may be caused by genetic mutations, epigenetic changes, or environmental factors, resulting in abnormal signaling that can promote disease progression (Gomperts et al., 2009).

Cancer is likely the best-documented disease linked to signal transduction pathway dysfunction. Many cancers are caused by mutations in genes that encode critical components of signaling pathways that control cell growth and survival. The Ras-Raf-MEK-ERK (MAPK) pathway, for example, is frequently misregulated in cancers such as melanoma, colorectal cancer, and lung (Rönnblom and Leonard, 2019). Mutations in Ras, a small GTPase that acts as an upstream activator of the MAPK pathway, are among the most common cancer-causing mutations. These mutations trap Ras in a constitutively active state, causing persistent activation of downstream kinases and driving uncontrolled cell division and tumor growth. Similarly, a kinase in the MAPK pathway, a kinase in the MAPK pathway mutations in BRAF, are prevalent in melanoma and other cancers, further highlighting the importance of signal transduction dysregulation in oncogenesis (Pollard et al., 2024).

The deregulation of the Akt/mTOR/phosphoinositide 3-kinase (PI3K) pathway, which is essential for controlling cell growth, survival, and metabolism, is another well-known example. Because of abnormalities in the genes encoding Akt, PTEN, or PI3K (a negative regulator of the system), this pathway is frequently hyperactivated in malignancies (Punekar, 2018). Tumor suppressor gene PTEN loss is very prevalent in malignancies like endometrial, prostate, and glioblastoma. The function of PTEN is to dephosphorylate phosphoinositides, which prevents Akt from being activated. Akt is still active when PTEN is absent or altered, which promotes cell growth and survival and aids in the formation of tumors (Yin and Shen, 2008). Aberrant PI3K/Akt pathway activation contributes to treatment resistance, as cancer cells with hyperactive Akt signaling can dodge apoptosis produced by chemotherapy or radiation (Rönnblom and Leonard, 2019).

Signal transduction dysregulation is also linked to neurodegenerative illnesses, in which faulty signaling causes progressive loss of neuronal function and structure. The Wnt/β-catenin signaling system impacts neuronal growth and synaptic plasticity. This pathway's dysregulation has been associated with Alzheimer's (AD) and Parkinson's (PD). Reduced Wnt signaling contributes to the production of amyloid plaques and tau tangles, which are hallmarks of Alzheimer's disease. Impaired Wnt signaling may diminish the protective effects of neurotoxins, resulting in neuronal death and cognitive impairment. Similarly, alterations in the Wnt pathway are likely to lead to dopaminergic neuron loss in the substantia nigra, a critical event in Parkinson's disease etiology (Jin et al., 2016).

Cardiovascular diseases (CVDs) are also strongly linked to signal transduction dysfunction. The Notch signaling pathway, which regulates vascular development and homeostasis, has been linked to atherosclerosis, hypertension, and heart failure. In atherosclerosis, faulty Notch signaling can impair endothelial cell function, causing inflammation, smooth muscle cell proliferation, and plaque development. Furthermore, hyperactivation of the PI3K/Akt pathway, which is protective in ischemia settings, can have negative effects if overactive in non-ischemic conditions, leading to pathological ventricular hypertrophy and heart failure (Gomperts et al., 2009). The complexity of these signaling networks in the cardiovascular system emphasizes the delicate balance that must be maintained to avoid illness.

Dysregulated signal transduction pathways also play a role in autoimmune illnesses such systemic lupus erythematosus (SLE) and rheumatoid arthritis. Overactivation of the Type I interferon (IFN) signaling pathway contributes significantly to the pathophysiology of SLE. This system typically functions as a defense mechanism against viral infections, but in SLE, it is continually engaged, resulting in chronic inflammation and tissue damage (Rönnblom and Leonard, 2019). Similarly, in RA, the disease's hallmarks of chronic inflammation and joint degradation are primarily driven by dysregulated signaling via the NF-κB pathway. Several scientific research has shown that, pro-inflammatory cytokines like TNF-α cause the NF-κB pathway to become activated, which in turn produces more cytokines and enzymes that break down joint tissues (Gomperts et al., 2009).

The significance of signal transduction pathways in metabolic illnesses, particularly diabetes, has also been thoroughly investigated. Insulin signaling via the PI3K/Akt pathway is critical for glucose homeostasis, controlling processes such as glucose absorption, glycogen synthesis, and lipid metabolism. Insulin resistance in type 2 diabetes is defined by poor signaling through this route, which results in reduced glucose uptake in muscle and adipose tissue and increased glucose synthesis by the liver (Barbu et al., 2020). Obesity, a key risk factor for type 2 diabetes, is linked to chronic inflammation and the release of cytokines like TNF-α and IL-6. This can disrupt insulin signaling by activating stress-related kinases that block insulin receptor signaling (Mollicone et al., 2009). The complex relationship between signal transduction and metabolic control highlights the possibility of targeting these pathways in treatment efforts for diabetes and other metabolic disorders.

To aid in their invasion, reproduction, and survival, pathogens frequently take over host cell signaling pathways in the context of infectious diseases. According to Dimitrov et al. (2017), viruses that cause influenza, hepatitis B and C, and HIV alter signaling pathways, such as NF-κB, PI3K/Akt, and MAPK in order to avoid the host immune response and spread the virus. For example, HIV impairs immunological signaling pathways that normally restrict viral replication while concurrently activating the NF-κB pathway to increase viral transcription. The ability of pathogens to disrupt host cell signaling pathways not only aids in their survival but also contributes to disease pathology, as chronic activation of these pathways can result in inflammation, tissue damage, and the development of cancers such as hepatocellular carcinoma in chronic hepatitis infection (Rönnblom and Leonard, 2019).

The potential for therapeutic approaches targeting dysregulated signal transduction pathways is enormous. Tyrosine kinase inhibitors (TKIs) have shown promise in treating cancers with specific signaling system alterations, such as EGFR mutations in lung cancer or BRAF mutations in melanoma (Sharma et al., 2007). Similarly, in autoimmune illnesses, biologics that target cytokines or their receptors, such as TNF inhibitors in rheumatoid arthritis, have transformed treatment by slowing dysregulated signaling and lowering inflammation. Given the prevalence of multiple signaling molecules in both healthy and sick tissues, discovering medicines that precisely control these pathways while preserving normal cellular function remains a problem.

Chapter **5**

DNA Replication and Repair

Contents

DNA replication and repair are critical processes for maintaining genomic integrity and accurate cell division. DNA replication is a highly regulated, complex process that involves unwinding the double helix, synthesizing a complementary strand, and ensuring fidelity through proofreading mechanisms. DNA polymerases, enzymes that synthesize new DNA strands by adding nucleotides in a sequence-specific manner, are fundamental to this process. These polymerases, especially DNA polymerase delta and epsilon in eukaryotes, have high fidelity due to their proofreading activity, which reduces errors during replication (Alberts et al., 2015).

Mismatches, single-strand breaks, and double-strand breaks are all examples of DNA damage that cells can repair through multiple pathways. The mismatch repair (MMR) system corrects errors overseen by DNA polymerases. The base excision repair (BER) and nucleotide excision repair (NER) pathways, respectively, address damage caused by reactive oxygen species and UV light. Double-strand breaks, which are one of the most lethal types of DNA damage, are primarily repaired by homologous recombination (HR) and non-homologous end joining (NHEJ). HR is a high-fidelity process that uses a sister chromatid as a template for repair, whereas NHEJ is faster but more error-prone (Pollard et al., 2022).

Understanding the molecular mechanisms of DNA replication and repair is critical for understanding the etiology of various diseases, including cancer, which is frequently associated with defects in these processes (Alberts et al., 2015).

5.1. DNA STRUCTURE AND FUNCTION

DNA is all living organisms' molecular blueprint, containing the instructions required for cellular function, development, and reproduction. DNA is a polymer made up of repeating units called nucleotides, which contain three components: a phosphate group, a deoxyribose sugar, and a nitrogenous base. Purines, which include adenine (A) and guanine (G), and pyrimidines, which include cytosine (C) and thymine (T), are the two types of nitrogenous bases (Alberts et al., 2015). These bases are critical for DNA's ability to store and transmit genetic information (Figure 5.1).

DNA's iconic structure, discovered by James Watson and Francis Crick in 1953, is a double helix made up of two antiparallel nucleotide strands. The strands are oriented in opposite directions, with one running 5' to 3' and the other 3' to 5.' This antiparallel arrangement is essential for complementary base-pairing between the strands, in which adenine forms hydrogen bonds with thymine and guanine pairs with cytosine (Pollard et al., 2022). These specific base-pairing rules ensure that the genetic code is precisely copied during DNA replication.

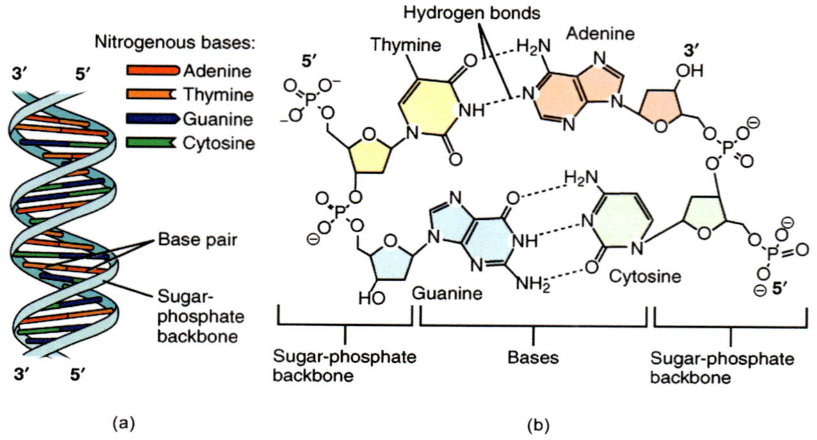

Figure 5.1. *Basic DNA structure.*

Source: https://courses.lumenlearning.net/biology/chapter/chapter-9-molecular-biology/.

The double helical structure of DNA is stabilized by a variety of forces, including hydrogen bonding between base pairs and hydrophobic interactions between stacked bases. The negatively charged phosphate backbone that interacts with the cell's aqueous environment protects the hydrophobic core formed by the stacked bases. This structural configuration not only ensures stability but also allows DNA to be compacted into chromatin within the nucleus of eukaryotic cells. Chromatin is further organized into higher-order structures, eventually forming chromosomes, which are required for proper genetic material segregation during cell division (Ransom et al., 2010).

The primary function of DNA is to store genetic information, which is encoded in the nitrogenous base sequences that run along its strands. This information is converted into messenger RNA (mRNA) through the transcription process, in which RNA polymerase creates a complementary RNA strand from the DNA template. The mRNA then transports genetic instructions from the nucleus to ribosomes in the cytoplasm, where they are translated into proteins. Proteins are the cell's workhorses, performing a wide range of functions such as enzymatic catalysis, structural support, and process regulation. The sequence of amino acids in a protein, which determines its structure and function, is directly determined by the sequence of bases in the corresponding gene (Branzei and Foiani, 2008).

Furthermore, DNA plays an important role in inheritance. During DNA replication, the double helix unwinds, with each strand serving as a template for the creation of a new complementary strand. This semi-conservative replication mechanism ensures that each daughter cell receives an exact copy of the genetic material. DNA replication fidelity is maintained by various proofreading mechanisms and repair systems that correct errors, preserving the genetic code's integrity over generations (Branzei and Foiani, 2008; Alberts et al., 2015).

Mutations, or changes in the DNA sequence, can result from replication errors, exposure to mutagens, or other causes. While some mutations may be neutral or even beneficial, others can have negative consequences, leading to diseases like cancer. The study of DNA mutations and their consequences is an important aspect of genetics and molecular biology, as it provides insights into evolutionary mechanisms and the development of genetic disorders.

In addition to storing and transmitting genetic information, DNA helps to regulate gene expression. Regulatory sequences within the DNA, such as promoters and enhancers, regulate the timing, location, and level of gene expression, ensuring that proteins are produced in the proper context. Epigenetic modifications, such as DNA methylation and histone acetylation, influence gene expression by changing the accessibility of the DNA to transcriptional machinery (Pollard et al., 2022). These regulatory mechanisms allow cells to respond to environmental cues, differentiate into different cell types, and maintain homeostasis.

5.2. DNA REPLICATION PROCESS

DNA replication is a fundamental process that ensures the correct transmission of genetic information from one generation to the next. This highly orchestrated process consists of a series of biochemical events that allow a cell to duplicate its DNA content prior to cell division. DNA replication is critical for growth, development, and genomic integrity (Alberts et al., 2015).

DNA replication begins at specific locations on the DNA molecule known as replication origins. These sites are distinguished by unique sequences that indicate the start of replication. In eukaryotes, each chromosome has multiple replication origins, allowing large genomes to be replicated simultaneously. Prokaryotic cells, on the other hand, have only one replication origin and typically have a single circular chromosome. The initiation phase is tightly controlled, ensuring that replication only happens once per cell cycle (Ransom et al., 2010; Pollard et al., 2022). The DNA double helix unwinds at the origin of replication, exposing the individual strands and forming the replication fork structure. Helicase, an enzyme that breaks the hydrogen bonds between complementary base pairs, aids in this unwinding process. The separation of the DNA strands results in a region of single-stranded DNA (ssDNA) that acts as a template for the synthesis of new DNA strands. However, because unwound ssDNA is prone to secondary structures and is susceptible to nucleases, it is stabilized by single-strand binding proteins (SSBs), which bind to the ssDNA and prevent it from reannealing or being degraded (Koutroumpas and Lygeros, 2011).

The enzyme DNA polymerase catalyzes DNA synthesis by adding nucleotides to the growing strand in a sequence-specific manner. DNA polymerase can only add nucleotides to the 3' end of an existing strand, so a primer is required to start synthesis. This primer is a short segment of RNA synthesized by the enzyme primase that contains the free 3' hydroxyl group required for DNA polymerase to begin adding deoxyribonucleotides. In eukaryotic cells, DNA polymerase α synthesizes the RNA-DNA hybrid primer, followed by elongation of the leading and lagging strands by DNA polymerase δ and ε (Alberts et al., 2015; Chagin et al., 2010).

The replication fork is asymmetrical because the DNA strands are antiparallel. One strand, known as the leading strand, is continuously synthesized in the same direction that the replication fork moves. The other strand, the lagging strand, is synthesized discontinuously in the opposite direction as a series of short fragments known as Okazaki fragments. Each Okazaki fragment requires a new RNA primer to initiate. As replication continues, the enzyme DNA ligase joins these fragments together, sealing the nicks in the sugar-phosphate backbone and forming a continuous strand (Pollard et al., 2022) (Figure 5.2). The accuracy of DNA replication is critical, as errors in the sequence can result in mutations, some of which are deleterious or even lethal. DNA polymerases have an intrinsic proofreading ability that significantly improves replication accuracy. This proofreading function is due to DNA polymerases' 3' to 5' exonuclease activity, which allows the enzyme to remove incorrectly incorporated nucleotides as soon as they are added. After removing the incorrect nucleotide, DNA polymerase re-inserts the correct nucleotide before proceeding with elongation.

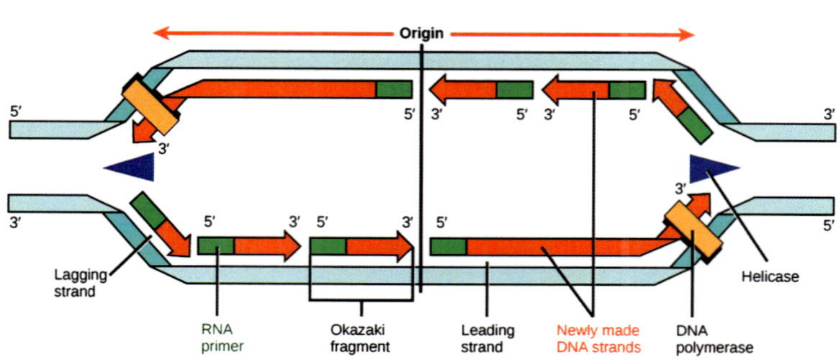

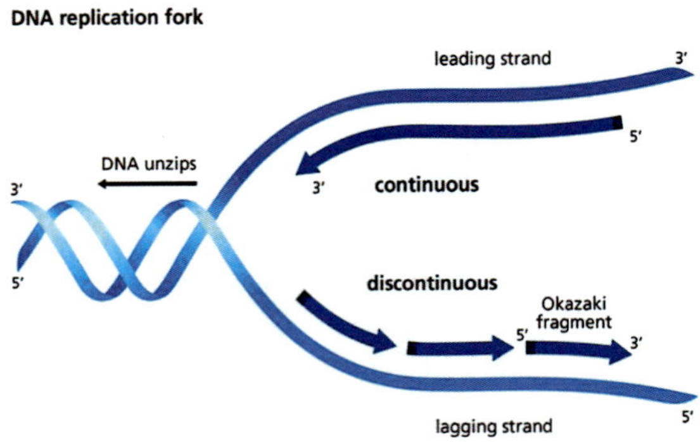

Figure 5.2. *Replication of leading and lagging strands of DNA.*

Source: https://courses.lumenlearning.net/biology/chapter/chapter-9-molecular-biology/.

This proofreading mechanism reduces the error rate to about one per billion nucleotides copied (Chagin et al., 2010; Koutroumpas and Lygeros, 2011).

In addition to proofreading, the cell has several DNA repair mechanisms that help to ensure the genome's integrity. Mismatch repair is one such mechanism that recognizes and corrects mismatched bases that escape DNA polymerase's proofreading activity. The mismatch repair system distinguishes between the newly synthesized strand and the parental strand based on nicks or methylation patterns, ensuring that the correct base is retained (Ransom et al., 2010).

Another important aspect of DNA replication is synchronization with the cell cycle. In eukaryotic cells, replication occurs during the S phase of the cell cycle, after a series of regulatory events that ensure the cell is ready for DNA synthesis. Cyclin-dependent kinases (CDKs) play an important role in this regulation by activating pre-replication complexes that form at replication origins during the G1 phase. These complexes are made up of various proteins, such as the origin recognition complex (ORC), Cdc6, Cdt1, and the minichromosome maintenance (MCM) helicase. When CDKs and other factors activate these complexes, DNA unwinding begins, and DNA polymerases and other replication machinery are recruited to the replication forks (Alberts et al., 2015; Pollard et al., 2022). As replication progresses, replication forks travel along the DNA, unwinding the helix and synthesizing new strands. However, the helical nature of DNA causes topological stress ahead of the replication fork, which results in the formation of supercoils. Topoisomerases, enzymes that break and rejoin DNA strands to relieve torsional strain, are responsible for the resolution of supercoils. Type I topoisomerases cause transient single-strand breaks, allowing the DNA to rotate and relax, whereas type II topoisomerases, such as DNA gyrase in prokaryotes, cause double-strand breaks to more effectively manage supercoiling (Pollard et al., 2022). After replication, the newly synthesized DNA strands must be separated from the parental strands. This process in prokaryotes involves topoisomerase IV-mediated decatenation of circular chromosomes. In eukaryotes, linear chromosome replication presents additional challenges, particularly at telomeres, the repetitive sequences at the ends of chromosomes. The end-replication problem arises when the lagging strand cannot be fully replicated, resulting in telomere shortening with each cell division (Chagin et al., 2010; Ransom et al., 2010). The enzyme telomerase counteracts this shortening by extending the telomeres with repetitive nucleotide sequences, preserving the integrity of the chromosome ends and preventing genomic instability.

5.3. DNA REPAIR MECHANISMS

DNA, the carrier of genetic information, is susceptible to a wide range of damaging agents that can alter its structure and sequence. If these damages are not repaired, they can cause mutations, genomic instability, and, eventually, diseases like cancer. To combat these threats, cells have developed a complex set of DNA repair mechanisms. These systems protect the genetic code by detecting and correcting various forms of DNA damage (Pollard et al., 2022) (Figure 5.3).

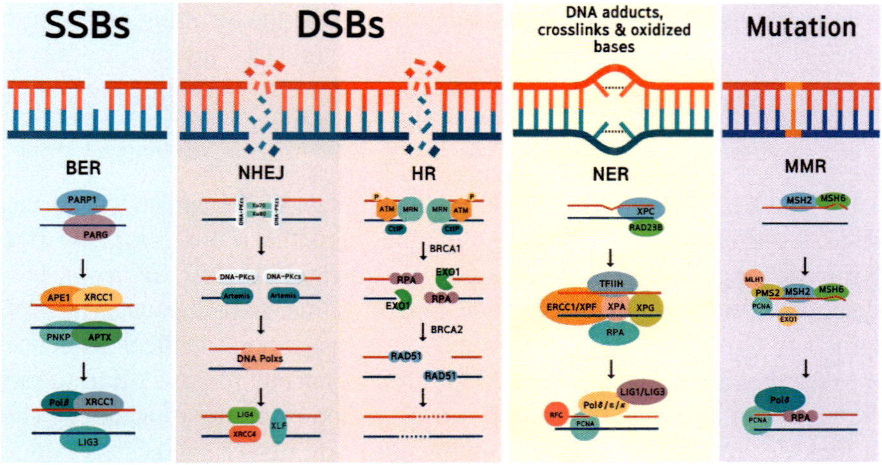

Figure 5.3. *DNA damage repaired by mismatch, base-excision, transcription-coupled/global genome, or homologous recombination/non-homologous end-joining mechanisms.*

Source: Moon J, Kitty I, Renata K, Qin S, Zhao F, Kim W. (2023). DNA damage and its role in cancer therapeutics. International Journal of Molecular Sciences 24(5), 4741. https://doi.org/10.3390/ijms24054741.

One of the primary DNA repair mechanisms is base excision repair (BER), which is in charge of correcting small, non-helix-distorting base lesions caused by oxidative stress, alkylation, deamination, or spontaneous hydrolysis. The process begins with a DNA glycosylase recognizing the damaged base and cleaving the N-glycosidic bond between the damaged base and the sugar-phosphate backbone, resulting in the formation of an abasic site (AP site). AP endonuclease recognizes this site and cleaves the phosphodiester backbone adjacent to it, resulting in a single-stranded gap. DNA polymerase then fills in the gap with the appropriate nucleotide, and DNA ligase seals the nick in the backbone, restoring the DNA to its original state. BER is essential for repairing endogenous DNA damage and preventing mutations that could disrupt cellular function (Alberts et al., 2015; Pollard et al., 2022).

Another important DNA repair pathway is nucleotide excision repair (NER), which addresses bulky, helix-distorting lesions caused by UV light-induced pyrimidine dimers and chemical adducts. NER has two sub-pathways: global genome NER (GG-NER) and transcription-coupled NER (TC-NER). GG-NER scans the entire genome for DNA damage, whereas TC-NER is activated only when RNA polymerase stalls at a lesion during transcription. In both pathways, a multi-protein complex recognizes the damage and recruits endonucleases to make incisions on both sides of the lesion, excising a short oligonucleotide that contains the damage. DNA polymerase then fills the resulting gap with the undamaged complementary strand as a template, and DNA ligase seals the nick to complete the repair. NER is particularly important for protecting the genome from UV radiation and environmental mutagens (Ransom et al., 2010; Furgason and Bahassi, 2013).

Mismatch repair (MMR) is a post-replication repair mechanism that corrects base-base mismatches and insertion-deletion loops that were missed by DNA polymerase's proofreading activity during replication. MMR improves DNA replication fidelity by identifying and repairing mismatches caused by DNA polymerase errors as well as recombination between non-identical sequences. The MutS protein complex recognizes the mismatch and recruits MutL to form a sliding clamp that searches for the nearest strand discrimination signal, such as a nick or an unsealed Okazaki fragment on the newly synthesized strand. Once the correct strand has been identified, an exonuclease is used to remove the mismatched nucleotides, and DNA polymerase resynthesizes the excised segment. Finally, DNA ligase seals the repaired segment to complete the process. MMR mutations are associated with microsatellite instability and an increased risk of certain cancers, particularly colorectal cancer (Liu et al., 2015).

Double-strand breaks (DSBs), one of the most severe types of DNA damage, can occur as a result of ionizing radiation, oxidative stress, and certain chemicals, as well as during normal cellular processes like meiotic recombination and V(D)J recombination in the immune system. Cells use two main pathways to repair DSBs: homologous recombination (HR) and non-homologous end joining (NHEJ). HR is an error-free repair mechanism that employs a homologous sequence, typically the sister chromatid, as a template to precisely repair the break. The process begins with resection of the DSB ends to produce single-stranded DNA coated with the Rad51 recombinase protein. Rad51 mediates the search for a homologous sequence and facilitates strand invasion, forming a displacement loop (D-loop) that pairs the damaged DNA with the template. DNA synthesis occurs with the intact template, and the resulting Holliday junctions are resolved, restoring DNA integrity. HR is limited to the S and G2 phases of the cell cycle when a sister chromatid is present (Alberts et al., 2015; Liu et al., 2015).

In contrast, NHEJ is an error-prone repair mechanism that directly ligates broken DNA ends without the use of a homologous template. NHEJ functions throughout the cell cycle, but it is especially important in the G1 phase. The process starts with the recognition and binding of the DSB and ends with the Ku70/80 heterodimer, which recruits DNA-PKcs to form the DNA-PK complex. This complex helps to align the broken ends and recruits additional factors, such as nucleases, polymerases, and ligases, to process and ligate them. While NHEJ is a fast and efficient repair mechanism, it can cause small insertions or deletions at the repair site, potentially resulting in mutations. Nonetheless, NHEJ is required for genomic stability, particularly in non-dividing cells, and for DSB repair during V(D)J recombination, a process critical for generating antibody diversity in the immune system (Pollard et al., 2022).

In addition to these major repair pathways, cells have specialized mechanisms for repairing specific types of damage. One such mechanism is direct repair, which can reverse certain types of DNA damage without the need for nucleotide excision or replacement. For example, the enzyme O6-methylguanine-DNA methyltransferase (MGMT) removes alkyl groups from guanine's O6 position, which can result in GC to AT transition mutations if not repaired. MGMT functions as a suicide enzyme, transferring an alkyl group to a cysteine residue in its active site and inactivating itself in the process. Direct repair is an important defense mechanism against alkylating agents, which are common environmental mutagens and chemotherapeutic agents (Furgason and Bahassi, 2013; Liu et al., 2015).

Another specialized repair mechanism is translesion synthesis (TLS), which allows DNA replication to continue after a lesion that would otherwise cause the replication fork to stall. TLS is performed by specialized DNA polymerases, like polymerase η, which can insert nucleotides opposite damaged bases with lower fidelity than replicative polymerases. While TLS enables cells to tolerate DNA damage, it is inherently mutagenic because the inserted nucleotides may not be correctly matched with the template. TLS is a last-resort mechanism that prevents replication fork collapse and cell death, but it increases mutagenesis (Branzei and Foiani, 2008).

Cells use checkpoint pathways to monitor DNA damage and coordinate repair with cell cycle progression. These pathways, mediated by proteins such as ATM, ATR, and Chk1/2, detect DNA damage and initiate a cascade of signaling events that halt the cell cycle, allowing time for repair before the cell moves on to the next phase. If the damage is too severe to repair, these pathways can cause programmed cell death (apoptosis) to stop the spread of damaged DNA (Alberts et al., 2015; Pollard et al., 2022).

5.4. TELOMERES AND TELOMERASE

Telomeres are specialized nucleoprotein structures that appear at the ends of linear chromosomes in eukaryotic cells. They are made up of repetitive DNA sequences, usually rich in guanine, and associated proteins that prevent chromosome ends from being recognized as DNA damage. In most vertebrates, the telomeric DNA sequence consists of thousands of repetitions of the hexanucleotide repeat TTAGGG. The primary function of telomeres is to maintain chromosomal stability by preventing the natural ends of chromosomes from being recognized as double-strand breaks (DSBs), which would otherwise result in inappropriate repair activities like non-homologous end joining (NHEJ) or homologous recombination (HR). These processes, if misdirected to telomeres, could lead to chromosomal fusions, genomic instability, and cell death (Alberts et al., 2015).

During DNA replication, the linear nature of eukaryotic chromosomes poses a unique challenge to the replication machinery, particularly at the terminal regions. To begin replication, the DNA polymerase enzyme complex, which is responsible for the synthesis of new DNA strands, requires an RNA primer. This requirement causes a problem with replication at the very ends of linear chromosomes, known as the "end-replication problem." Once the RNA primer at the 5' end of the lagging strand is removed, there is no upstream DNA segment to provide a free 3'-OH group for DNA polymerase to extend from, resulting in telomeres gradually shortening with each cell division (Shay and Wright, 2019).

Telomerase, a ribonucleoprotein enzyme complex, is critical in counteracting telomere shortening. Telomerase is composed of two core components: the telomerase reverse transcriptase (TERT) protein and an RNA component known as telomerase RNA (TERC), which acts as a template for the addition of telomeric repeats to the chromosome's 3' end. TERT catalyzes telomere extension by synthesizing new DNA from the TERC template, replenishing telomeric sequences lost during replication. This extension process compensates for the inevitable loss of telomeric DNA and allows cells to avoid replicative senescence, which is characterized by irreversible growth arrest caused by critically short telomeres (Pfeiffer and Lingner, 2013).

Telomerase activity in cells is tightly controlled. Telomerase is highly active in embryonic stem cells, germ cells, and some adult stem cells, but it has little to no activity in most somatic cells. As a result, somatic cells gradually shorten their telomeres with each cell division, acting as a molecular clock that limits cell proliferation—a phenomenon known as the Hayflick limit. When telomeres become critically short, they lose their protective function, triggering DNA damage responses (DDRs). These responses can cause cell cycle arrest through the tumor suppressor proteins p53 and Rb, or they can cause apoptosis if the damage is irreversible (Pollard et al., 2022). Telomere shortening acts as a tumor-suppressive mechanism by limiting cell lifespan and preventing uncontrolled cell proliferation.

Telomerase activity is, however, reactivated in certain cell types, particularly cancer cells, allowing these cells to maintain telomere length and proliferate indefinitely. The reactivation of telomerase is a hallmark of cancer, contributing to cancer cells' immortality by avoiding the normal replicative senescence caused by telomere shortening. Telomerase activity is upregulated in approximately 85-90% of human cancers, making it a promising target for cancer therapies. Telomerase inhibitors are being investigated as potential therapeutic agents for limiting cancer cell replication by inducing telomere shortening and subsequent cell death (Shay and Wright, 2019).

In addition to telomere maintenance, telomerase has been linked to other cellular processes such as DNA repair and gene expression regulation. Telomerase is thought to assist with DSB repair by facilitating the processing of broken DNA ends, but this role is not fully understood. Furthermore, emerging evidence indicates that telomerase may influence gene expression in addition to its telomere-lengthening function, potentially affecting cellular signaling pathways, differentiation, and survival (Pfeiffer and Lingner, 2013).

Telomere length and telomerase activity are regulated in a complex manner with multiple layers of control, including TERT transcriptional regulation, post-transcriptional modifications, alternative splicing of TERT mRNA, and TERC availability. Telomeric proteins, such as shelterin, a multiprotein complex that binds specifically to telomeric DNA, also play an important role in regulating telomere maintenance. Shelterin components such as TRF1, TRF2, POT1, and

TIN2 protect telomeres from being identified as DNA damage and regulate telomerase access to telomere ends. TRF2, for example, is essential for the formation of the T-loop structure, in which the 3' overhang of the telomere invades the duplex telomeric DNA, thus sequestering the chromosome end and further preventing unwanted DNA repair activities (Pfeiffer and Lingner, 2013; Pollard et al., 2022).

Telomere maintenance mechanisms can be disrupted, resulting in a variety of diseases known as telomere syndromes or telomeropathies. These disorders are distinguished by premature telomere shortening and include dyskeratosis congenita, idiopathic pulmonary fibrosis, and aplastic anemia. In these diseases, mutations in genes encoding components of the telomerase complex or shelterin proteins result in defective telomere maintenance, leading to genomic instability, stem cell exhaustion, and impaired tissue regeneration (Alberts et al., 2015).

5.5. MUTATIONS AND GENETIC DISEASES

Mutations are crucial to the study of genetic illnesses in the field of cell biochemistry, where knowing the molecular mechanisms driving genetic changes is critical for determining disease etiology. At its most fundamental, a mutation is any heritable change in the DNA sequence. Such alterations can occur spontaneously as a result of mistakes in DNA replication or can be caused by external causes such as radiation, chemical exposure, or viral infections. The molecular repercussions of mutations are determined by the type of change and its position within the genome. When a mutation arises in a gene that codes for a protein, it might affect the structure or function of that protein, potentially causing disease (Sclafani and Holzen, 2007). Understanding how these mutations manifest as disease phenotypes necessitates a thorough understanding of the gene's molecular biology as well as the biochemical properties of the proteins it encodes.

Mutations can be classified into numerous groups based on the type of genetic change they reflect. Point mutations, which entail the substitution of one nucleotide for another, are the most basic type of mutation, but they can have a considerable impact on protein functionality. For example, a point mutation that results in the replacement of one amino acid for another in a crucial area of a protein can affect its folding, stability, or capacity to interact with other molecules, potentially leading to a loss or gain of function (Alberts et al., 2015). Missense mutations, which substitute one amino acid with another, and nonsense mutations, which add a premature stop codon, can have a significant impact on the synthesis of protein and function. Nonsense mutations, in particular, frequently result in shortened, non-functional proteins, as seen in illnesses such as Duchenne muscular dystrophy, in which the absence of a functioning dystrophin protein causes progressive muscle degeneration (Kouzarides, 2007).

Insertions and deletions are two other types of mutations that can have a far greater impact on gene function. These mutations can disrupt a gene's reading frame, resulting in abnormal protein synthesis. Frameshift mutations, which arise when nucleotides are added or deleted in non-multiples of three, cause the genetic code's triplet reading frame to shift, resulting in premature protein synthesis. Such mutations are commonly seen in genetic diseases like cystic fibrosis, where deletion of three nucleotides in the CFTR gene results in the loss of a single phenylalanine residue, severely impairing the function of the CFTR protein, a chloride channel involved in the regulation of salt and water balance in epithelial tissues (Baynes and Dominiczak, 2009).

Enzymopathies, a type of genetic disease caused by enzyme abnormalities, provide a particularly good example of the biochemical consequences of mutations. Enzymes are biological catalysts that speed up chemical reactions in cells; even minor structural changes caused by genetic mutations can impair their catalytic activity. In many situations, these mutations result in a loss of function, such as in phenylketonuria (PKU), a condition caused by mutations in the PAH gene, which produces the enzyme phenylalanine hydroxylase. This enzyme typically converts the amino acid phenylalanine to tyrosine, but in people with PKU, mutations in PAH limit the enzyme's activity, resulting in hazardous quantities of phenylalanine in the blood and, if left untreated, significant intellectual disability (Baker, 2009). Such mutations frequently alter biochemical pathways, resulting in an accumulation of substrates or a lack of products, which can have far-reaching consequences for cellular metabolism and functionality.

Not all mutations result in the full loss of protein function; certain mutations might result in a gain of function, in which an altered protein exhibits new or improved activity. Such mutations are widespread in oncogenes, where they can cause uncontrolled cell growth and proliferation, which is characteristic of cancer. For example, mutations in the RAS family of genes, which encode small GTPases important in cell signaling, can trap the RAS protein in an active, GTP-bound state, resulting in the continual activation of downstream signaling pathways that promote cell proliferation and survival (Sclafani and Holzen, 2007). The gain-of-function mutations in RAS are among the most common genetic alterations in human cancers, indicating the importance of understanding the biochemical consequences of mutations in disease contexts.

In addition to small insertions and point mutations or deletions, larger-scale chromosomal alterations can have a significant impact on gene expression and functionality. These alterations can include deletions, translocations, inversions, and duplication of large DNA sequences. Chromosomal mutations frequently influence numerous genes simultaneously, resulting in complicated disease manifestations. For example, Charcot-Marie-Tooth disease, a hereditary neuropathy, is frequently caused by the duplication of a significant stretch of DNA on chromosome 17, including the PMP22 gene. Overexpression of the PMP22 protein impairs peripheral nerve function, resulting in gradual muscular weakening and sensory loss (Sclafani and Holzen, 2007).

Mutations have an impact on gene function and disease that extends beyond protein-coding genes. Mutations in non-coding sections of the genome, such as promoters, enhancers, and introns, can alter gene control and cause disease. Mutations in the promoter region of the TERT gene, which encodes the catalytic subunit of the enzyme telomerase, have been implicated in several malignancies (Furgason and Bahassi, 2013). These mutations improve TERT expression, allowing cancer cells to keep their telomeres intact and divide indefinitely. Similarly, mutations in splicing sites inside introns might cause pre-mRNA to be misprocessed, resulting in the creation of aberrant transcripts. Such splicing abnormalities are a common cause of genetic illnesses, including spinal muscular atrophy, in which mutations in the SMN1 gene result in the loss of functional SMN protein, causing motor neuron degeneration (Furgason and Bahassi, 2013; Liu et al., 2015).

The biochemical processes underpinning genetic disorders also include the phenomena of genetic anticipation, in which the severity of a disease advances while the age of onset decreases in subsequent generations. This behavior is frequently observed in trinucleotide repeat expansion illnesses such as Huntington's disease and fragile X syndrome. These illnesses are caused by the proliferation of repetitive DNA sequences within or near genes, resulting in the generation of mutant proteins with hazardous characteristics. In Huntington's disease, for example, the amplification of a CAG repeat in the HTT gene produces a protein with an excessively long polyglutamine tract, which collects in neurons and causes neurodegeneration (Liu et al., 2015). The length of the repeat expansion correlates with the severity of the disease, and the tendency for the repeat to expand further in subsequent generations accounts for the phenomenon of anticipation.

Advances in biochemistry have shed light on the significance of epigenetic alterations in disease. Epigenetics refers to changes in gene expression that do not entail DNA sequence changes, but rather modifications to the DNA molecule or the histone proteins that surround DNA. These changes can include adding methyl groups to DNA or acetylating histones, both of which can affect gene expression by changing DNA's accessibility to the transcriptional machinery. Epigenetic alterations, such as abnormal DNA methylation patterns, have been linked to a variety of disorders, including cancer, where hypermethylation of tumor suppressor genes can inhibit their expression and contribute to carcinogenesis (Sclafani and Holzen, 2007).

5.6. REGULATION OF DNA REPLICATION

DNA replication is a closely regulated and highly coordinated process in cell biochemistry that is required to keep the genome intact during cell division. The regulation of DNA replication guarantees that the complete genome is duplicated precisely and only once every cell cycle. Replication errors can cause mutations, genomic instability, and possibly malignant shifts, therefore precisely controlling this process is crucial for cellular homeostasis (Leonard and Mechali, 2013). DNA replication begins in certain places of the genome known as replication origins. In eukaryotic cells, replication is controlled by a complex interaction of proteins that ensures replication happens at the right moment during the cell cycle, primarily during the S phase, and replication origins fire only once per cycle to prevent re-replication.

During the G1 phase of the cell cycle, the pre-replicative complex (pre-RC) is assembled at the replication origins, controlling the start of DNA replication. The first step in this process is the binding of the origin recognition complex (ORC) to the DNA, which acts as a landing pad for other components needed to start replication. The MCM (minichromosome maintenance) helicase complex is recruited by the ORC via the helicase-loading proteins Cdc6 and Cdt1. To facilitate the synthesis of new DNA strands, the MCM complex, a hexameric helicase, unwinds the DNA double helix (Deegan and Diffley, 2016). The pre-RC is built but not yet functioning because the MCM helicase is dormant during the G1 phase.

DNA replication is activated during the G1 to S phase and is controlled by Dbf4-dependent kinase (DDK) and cyclin-dependent kinases (CDKs). Cyclin proteins, which are generated and destroyed in a manner depending on the cell cycle, tightly regulate the activity of CDKs, which are important regulators of the cell cycle (Jin et al., 2016). Early in the S phase, the MCM complex and other pre-RC components like Sld2 and Sld3 are phosphorylated by CDKs and DDK, which makes it easier to recruit more proteins needed for the start of replication. Among these proteins are Cdc45 and the GINS complex, which, along with the MCM helicase, form the CMG (Cdc45-MCM-GINS) complex, the active helicase responsible for unwinding DNA at replication forks (Pollard et al., 2024). The phosphorylation-dependent activation of the MCM helicase is an important regulatory step since it commits the cell to initiate DNA replication. The activity of several checkpoint mechanisms, which keep a close watch on

DNA integrity and make sure replication is synchronized with other cell cycle events, further controls the replication of DNA. The S-phase checkpoint is one of the most well-studied checkpoints; it reacts to replication stress, such as DNA damage or stalled replication forks, and stops cell cycle progression or further origin firing until the issue is fixed. The buildup of single-stranded DNA, which is bound by the replication protein A (RPA) complex, triggers the S-phase checkpoint. The ATR (ATM and Rad3-related) kinase is brought in by this complex, and it phosphorylates and activates the Chk1 kinase, which in turn inhibits CDKs and stabilizes replication forks that have stalled (Kouzarides, 2007). To protect the fidelity of the replication process, this checkpoint makes sure that DNA replication only starts when the template is intact and that it is halted until any damage is fixed.

Apart from these checkpoint mechanisms, the chromatin environment also plays a role in regulating DNA replication. DNA and histone proteins make up chromatin, which needs to be changed to give the replication machinery access. The acetylation of histone proteins, which is connected to chromatin opening and enhanced DNA accessibility, is one significant regulatory mechanism (Leonard and Mechali, 2013). To neutralize the positive charge on histone tails and decrease their attraction for DNA, histone acetyltransferases (HATs) catalyze the addition of acetyl groups, which makes it easier for replication factors to bind to DNA. On the other hand, acetyl groups are eliminated by histone deacetylases (HDACs), which encourage chromatin condensation and lessen DNA's accessibility to the replication machinery. The control of replication origin firing and replication fork advancement depends critically on the dynamic balance between histone acetylation and deacetylation.

Moreover, transcription and repair, two more biological functions, need to be coordinated with DNA replication. The control of replication time is essential because a conflict between the transcription and replication machinery might result in genomic instability. Replication timing is regulated in eukaryotic cells so that gene-poor and heterochromatic regions duplicate later in the S phase, whereas portions of the genome rich in actively transcribed genes prefer to replicate early. Replication fork stalling and DNA double-strand breaks can result from collisions between the replication and transcription machinery, which can be reduced by the temporal control of replication. In quickly proliferating cells, such as cancerous or embryonic cells, where the transcription and replication programs are highly active and strictly controlled to preserve genomic stability, this coordination is especially crucial (Leonard and Mechali, 2013).

Replication fork stability is another important element of DNA replication regulation. The replication fork is the structure formed when the DNA double helix is unwound and both strands are copied at the same time. Maintaining the stability of the replication fork is crucial to ensure that replication runs smoothly and properly. Several proteins, including those of the replication fork protection complex, stabilize the replication fork by inhibiting the dissociation of the replisome, the multi-protein complex in charge of DNA synthesis. If the replication fork collapses or the replisome dissociates, the cell must repair the damage before continuing replication. This repair is accomplished through homologous recombination, which employs a sister chromatid as a template to repair broken DNA and restore replication fork stability (Deegan and Diffley, 2016).

To achieve complete genome duplication, the termination of DNA replication must also be precisely regulated. Replication forks converge and end at specified points in eukaryotic cells, most commonly in repeated DNA sequences known as telomeres. Telomeres are specialized structures at the ends of chromosomes made up of repeated DNA sequences and related proteins. Because normal DNA polymerases cannot fully copy the ends of linear chromosomes, the enzyme telomerase extends the telomeres to prevent them from shortening during each replication cycle (Deegan and Diffley, 2016). Telomerase activity must be regulated to preserve genome integrity, as excessive telomere shortening can result in chromosome fusion and genomic instability, whereas overactive telomerase can contribute to oncogenesis by allowing cells to avoid senescence and cycle indefinitely.

Preventing re-replication is an important component of DNA replication regulation since it can lead to genomic instability and polyploidy, a state in which cells have more chromosomes than usual. Several systems inhibit re-replication by limiting replication origin licensing to once per cell cycle. One such mechanism is that increased CDK activity inhibits pre-RC formation during the S, G2, and M phases (Pfeiffer and Lingner, 2013). CDK-mediated phosphorylation of pre-RC components, including ORC, Cdc6, and MCM, prevents their re-association with replication origins, guaranteeing that no further replication initiation occurs after DNA replication begins. This tightly regulated inhibition of pre-RC assembly ensures that each segment of the genome is replicated only once per cell cycle, maintaining the proper genomic content in daughter cells.

The existence of replication origins, which differ in their activation timing and efficiency, further affects the regulation of DNA replication. While some origins are "inefficient" and only fire in certain situations or in reaction to replication stress, others are regarded as "efficient" and are often employed to start replication (Kouzarides, 2007). Cells are able to adapt to varying cellular states and external circumstances through the control of origin efficiency and timing, which guarantees that replication occurs under ideal circumstances.

Because dormant origins can be activated to finish replication when needed, the flexibility in origin usage also gives the cell a way to deal with DNA damage or replication fork stalling.

5.7. APPLICATIONS OF DNA REPLICATION IN BIOTECHNOLOGY

DNA replication has been essential in the advancement of biotechnology, with the basic principles of this natural process being used for a variety of applications in research, medicine, and industry. In cell biochemistry, the ability to replicate DNA in vitro and alter the replication process has transformed molecular biology, resulting in advances in diagnostics, treatments, genetic engineering, and synthetic biology. These applications use DNA replication mechanisms to amplify, change, and evaluate genetic material, giving valuable tools for biotechnological breakthroughs (Pollard et al., 2024).

One of the most well-known uses of DNA replication in biotechnology is the polymerase chain reaction (PCR), which allows for the amplification of specific DNA sequences. PCR uses DNA polymerase, the enzyme responsible for generating new DNA strands during replication, to repeat a target sequence in vitro. The process requires repeated rounds of denaturation, primer annealing, and extension of the DNA strand by the polymerase, which results in the exponential amplification of the target sequence (Jin et al., 2016). Because of its versatility and sensitivity, PCR has emerged as a key component of modern biotechnology. It is commonly employed in molecular diagnostics to detect pathogens such as viruses and bacteria, as well as genetic alterations linked to cancer and hereditary disorders (Van Elden et al., 2001). PCR is also used in forensic science for DNA profiling, which allows individuals to be identified based on their unique genetic composition (Baynes and Dominiczak, 2009).

The invention of PCR has also aided progress in genetic engineering, where DNA sequence amplification and modification are critical for the manufacture of recombinant proteins, gene therapy vectors, and genetically modified organisms (GMOs). PCR enables scientists to clone specific DNA sequences into expression vectors, which may then be injected into host cells to create recombinant proteins. These proteins have a broad range of uses, from industrial enzymes used in detergents and food processing to medicinal proteins such as insulin and monoclonal antibodies. In gene therapy, PCR is used to create therapeutic genes that can be given to patients to fix genetic flaws or treat diseases including cystic fibrosis and muscular dystrophy (Deegan and Diffley, 2016). Another key application of DNA replication in biotechnology is DNA sequencing, which establishes the exact order of nucleotides in a DNA molecule. DNA sequencing technologies, such as Sanger sequencing and next-generation sequencing (NGS), use DNA polymerases to synthesize new strands of DNA based on a template, with labeled nucleotides added to allow each base to be detected and identified (Pfeiffer and Lingner, 2013; Pollard et al., 2022). NGS, in particular, has transformed genomics by allowing for the quick and cost-effective sequencing of whole genomes. This technique has been crucial in promoting personalized medicine by using individuals' genetic information to drive diagnosis, treatment, and preventative methods. NGS is also utilized in agricultural biotechnology to improve crop species through genome editing and marker-assisted breeding, allowing for the generation of plants with improved features like drought tolerance, pest resistance, and yield (Punekar, 2018).

The ability to replicate and alter DNA has also resulted in the development of CRISPR-Cas9, a groundbreaking gene-editing technique that enables precise genome modifications. CRISPR-Cas9 is based on a naturally occurring bacterial immune system that guides RNAs to target certain DNA regions for cleavage by the Cas9 enzyme (Deegan and Diffley, 2016). Following cleavage, the cell's DNA repair processes are used to introduce the desired mutations or repairs to the target spot. While CRISPR-Cas9 is not directly tied to the process of DNA replication, it uses the cell's natural DNA repair mechanism to achieve gene editing, which is fundamentally linked to the replication and maintenance of the genome. This approach has enormous potential in biotechnology, with applications ranging from functional genomics and drug discovery to gene therapy and genetically modified organism production (Jinek et al., 2012). For example, CRISPR-Cas9 is being investigated for the treatment of genetic illnesses such as sickle cell anemia, where it can be used to fix the mutant gene that causes the disease.

In addition to its involvement in gene editing, DNA replication is used in synthetic biology, which aims to design and build novel biological systems and functionalities. Synthetic biology frequently entails the creation of artificial gene circuits and the synthesis of complete genomes, which necessitates the replication and assembly of massive DNA molecules. One famous example of synthetic biology is the construction of synthetic organisms with minimal genomes, such as *Mycoplasma laboratorium*, a bacterium whose genome was built from scratch and engineered to contain only the basic genes needed for life (Jin et al., 2016). Advances in DNA replication technology, such as the ability to reliably copy and assemble vast sections of DNA, enable the creation of such minimum genomes. Synthetic biology has the potential to transform industries such as biofuels, pharmaceuticals, and agriculture by allowing the creation of novel organisms with specific functions (Shay and Wright, 2019). DNA replication is used in materials science and nanotechnology in addition to genetic engineering and synthetic biology. Because base-pairing interactions allow DNA to generate exact, predictable structures, DNA is used as a building block for nanoscale electronics and structures. For instance, the craft of "DNA origami" employs the concepts of DNA folding and replication to produce intricate three-dimensional nanostructures (Gomperts et al., 2009). These structures have potential implications in medication delivery, as DNA nanostructures can be engineered to transport therapeutic chemicals and release them at precise body sites. DNA-based materials are also being investigated for application in biosensing and diagnostics because of their programmability, which allows them to identify specific chemicals or infections.

The manipulation of DNA replication has also resulted in breakthroughs in diagnostics, particularly the development of isothermal amplification techniques like loop-mediated isothermal amplification (LAMP) and recombinase polymerase amplification (RPA). Unlike PCR, which requires heat cycling, these techniques enable DNA amplification at a constant temperature, making them excellent for point-of-care diagnostics in resource-limited settings. LAMP, for example, is used to detect infectious disorders such as malaria, tuberculosis, and COVID-19, allowing for rapid and reliable diagnosis without the requirement of costly laboratory equipment (Li et al., 2019). The ability to duplicate DNA under simplified settings has significantly increased the availability of molecular diagnostics, particularly in countries with poor healthcare infrastructure.

DNA replication is also involved in the development of DNA vaccines, a new technique with potential uses in infectious illnesses and cancer treatment. DNA vaccines function by inserting a plasmid containing a gene coding for an antigen into host cells, where it is duplicated and expressed. This causes an immunological response to the antigen, granting immunity against the disease or malignancy (Li et al., 2019). DNA vaccines have various advantages over traditional vaccinations, including the capacity to stimulate both humoral and cellular immunity, as well as their durability and ease of manufacture. Recent advancements in DNA replication technology have enhanced the efficiency of plasmid replication and expression, increasing the efficacy of DNA vaccines in clinical studies (Shay and Wright, 2019).

Chapter 6

Gene Expression and Regulation

Contents

Gene expression and regulation are critical processes that govern the conversion of genetic information within DNA into functional products, primarily proteins. Gene expression is tightly controlled by a complex interplay of transcriptional, post-transcriptional, translational, and post-translational mechanisms that ensure genes are expressed at the appropriate time, location, and quantity (Baynes and Dominiczak, 2009).

Transcription factors, which bind to specific DNA sequences such as promoters and enhancers, regulate gene expression by activating or repressing transcription. Epigenetic modifications, such as DNA methylation and histone modifications, have a further impact on gene accessibility and transcriptional activity, adding an extra layer of regulation. Post-transcriptional RNA processing events such as splicing, editing, and polyadenylation affect the stability and translation efficiency of mRNA molecules. Non-coding RNAs, such as microRNAs, also play an important role by binding to target mRNAs and preventing translation or promoting degradation (Pollard et al., 2022).

Translational regulation is the process of controlling protein synthesis from mRNA in response to cellular signals or environmental conditions. Finally, post-translational modifications such as phosphorylation, ubiquitination, and glycosylation regulate protein activity, localization, and stability, fine-tuning their functions (Alberts et al., 2015).

These regulatory mechanisms are critical for the maintenance of cellular homeostasis, development, and response to external stimuli. Aberrations in gene regulation have been linked to a variety of diseases, including cancer, neurodegenerative disorders, and metabolic syndromes, emphasizing the importance of molecular understanding of these processes (Alberts et al., 2015).

6.1. TRANSCRIPTION AND TRANSLATION

Transcription commences with the initiation phase, in which the enzyme RNA polymerase binds to a specific region of DNA called the promoter. This promoter region is located upstream of the gene that will be transcribed and contains specific sequences that transcription factors recognize, such as the TATA box in eukaryotes. These transcription factors play an important role in recruiting RNA polymerase to the proper initiation site, ensuring that transcription occurs at the appropriate time and location. The binding of RNA polymerase causes the DNA to unwind, revealing the template strand, which serves as the blueprint for RNA synthesis (Alberts et al., 2015; Pollard et al., 2022) (Figure 6.1).

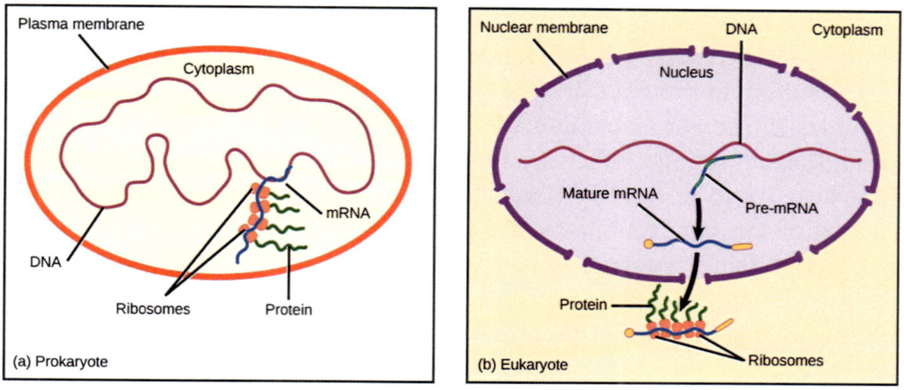

Figure 6.1. *Gene expression: A vital process in cells that ensures specific proteins are created at the right time and in the right quantities.*

Source: https://courses.lumenlearning.com/suny-wmopen-biology1/chapter/regulation-of-gene-expression/.

Once bound, RNA polymerase starts synthesizing a single-stranded RNA molecule by adding ribonucleotides that are complementary to the DNA template. This synthesis proceeds in the 5' to 3' direction, with the RNA strand elongating as the polymerase moves across the DNA template. The process is tightly regulated, with several factors influencing the speed and accuracy of RNA synthesis. Elongation factors and pausing and backtracking mechanisms help RNA polymerase move efficiently along DNA and correct errors in the sequence (Ramírez-Clavijo and Montoya-Ortíz, 2013).

The transcription process culminates when the RNA polymerase releases the newly synthesized RNA molecule, known as pre-mRNA in eukaryotes or mRNA in prokaryotes. Termination in prokaryotes frequently occurs through the formation of a hairpin structure in the RNA or the action of the Rho protein, which disrupts the interaction between RNA polymerase and the DNA template. In eukaryotes, termination is more complex, requiring the cleavage of the RNA transcript and the addition of a poly-A tail, which is critical for RNA stability and export from the nucleus (Kornblihtt et al., 2013).

In eukaryotic cells, pre-mRNA is subjected to extensive processing before being translated into a protein. This processing includes intron removal via splicing, which is done by the spliceosome, a complex of small nuclear RNAs and proteins. Exons, or the gene's coding regions, are joined together to form the mature mRNA. A 5' cap is also added to the RNA, which protects it from degradation and is required for the initiation of translation. The mature mRNA is then transported from the nucleus to the cytoplasm, where it acts as a template for protein synthesis.

The next stage of gene expression is translation, which is the process of decoding the sequence of nucleotides in mRNA into a sequence of amino acids, resulting in protein synthesis. This process takes place in the ribosome, a complex molecular machine made up of rRNA and proteins. The ribosome reads mRNA in codons, which are sets of three nucleotides that each specify an amino acid. Transfer RNAs (tRNAs) play an important role in this process by transporting the required amino acids to the ribosome. Each tRNA molecule has an anticodon that is complementary to a specific mRNA codon, allowing it to bind with the mRNA and ensure the correct amino acid is incorporated into the growing polypeptide chain (Ramírez-Clavijo and Montoya-Ortíz, 2013; Alberts et al., 2015).

There are three stages to the translation process: initiation, elongation, and termination. Initiation begins when the ribosome's small subunit binds to the mRNA near the start codon, usually AUG, which codes for methionine. Initiation factors facilitate this step, which requires energy in the form of GTP. The initiator tRNA, which contains methionine, pairs with the start codon, and the large ribosomal subunit joins the complex, forming the complete ribosome ready for elongation (Pollard et al., 2022).

During elongation, the ribosome moves along the mRNA, decoding each codon and adding its corresponding amino acid to the growing polypeptide chain. This process is extremely dynamic, requiring the coordination of tRNA entry into the ribosome, peptide bond formation, and ribosome translocation along the mRNA. Elongation factors help with the proper placement of tRNAs and the movement of the ribosome, ensuring efficient and accurate translation. The ribosome's peptidyl transferase activity catalyzes the formation of peptide bonds between adjacent amino acids, thus extending the polypeptide chain (Baynes and Dominiczak, 2009).

When the ribosome comes across a stop codon, such as UAA, UAG, or UGA, translation stops. These codons do not correspond to any tRNA; rather, release factors recognize them, causing the ribosome to release the newly synthesized polypeptide. The ribosome then dissociates into its subunits, preparing to initiate another round of translation on a different mRNA (Alberts et al., 2015).

The newly synthesized protein is then subjected to further modifications before becoming fully functional. These post-translational modifications may include the addition of chemical groups, folding into the proper three-dimensional structure, or cleavage of specific sequences to activate the protein. Chaperone proteins help the polypeptide fold properly, preventing misfolding and aggregation, both of which can be harmful to cellular function. In other cases, proteins may be targeted to specific cellular locations, such as the nucleus, mitochondria, or cell membrane, to perform their intended function (Alberts et al., 2015; Ramírez-Clavijo and Montoya-Ortíz, 2013).

6.2. RNA PROCESSING AND MODIFICATION

RNA processing and modification are critical stages in the maturation of RNA molecules, converting primary transcripts into functional RNA species that participate in a variety of cellular processes. These steps are critical for the proper expression of genetic information and the maintenance of cellular homeostasis, reflecting the complex mechanisms that have evolved to regulate gene expression post-transcriptionally (Baynes and Dominiczak, 2009).

In eukaryotic cells, RNA processing begins immediately after transcription, with the nascent RNA, or pre-mRNA, undergoing a series of modifications to prepare it for translation. One of the first modifications is to add a 7-methylguanosine cap to the 5' end of the pre-mRNA. A series of enzymes, including RNA triphosphatase, guanylyltransferase, and methyltransferase, catalyzes the 5' capping process by sequentially attaching the cap structure. The 5' cap serves several functions, including protecting the RNA from exonuclease degradation, facilitating nuclear export, and is required for translation initiation. Furthermore, the cap structure helps the ribosome recognize the mRNA, emphasizing its importance in gene expression (Pollard et al., 2022).

Following capping, the pre-mRNA is spliced, which removes introns (non-coding regions) and joins exons (coding regions) into a continuous coding sequence. The spliceosome, a large ribonucleoprotein complex made up of small nuclear RNAs (snRNAs) and associated proteins, is responsible for splicing. The spliceosome recognizes specific sequences at intron-exon boundaries, known as splice sites, and catalyzes the removal of introns using two transesterification reactions. Splicing regulation must be precise in order to produce mature mRNAs with the correct coding sequences. Furthermore, alternative splicing, a process in which different combinations of exons are joined together, enables the production of multiple protein isoforms from a single gene, greatly expanding the proteome's diversity (Licatalosi and Darnell, 2010; Alberts et al., 2015).

The splicing process is closely linked to transcription because spliceosome components are recruited to the pre-mRNA during its synthesis. This coupling ensures that RNA is processed in a timely and accurate manner, preventing the accumulation of unspliced or improperly spliced transcripts that could harm the cell. Furthermore, the splicing machinery interacts with the transcription machinery, influencing both splice site selection and splicing efficiency. This interaction between transcription and splicing represents the coordinated regulation of gene expression at multiple levels (Licatalosi and Darnell, 2010). Following splicing, the pre-mRNA goes through 3' end processing, which includes RNA cleavage downstream of a polyadenylation signal and the addition of a polyadenine (poly-A) tail to the 3' end. A complex of proteins, including cleavage and polyadenylation specificity factor (CPSF), cleavage stimulation factor (CstF), and poly-A polymerase, facilitates the polyadenylation process. The poly-A tail increases mRNA stability by protecting it from exonucleolytic degradation and is important in regulating mRNA translation and degradation. Furthermore, the length of the poly-A tail can affect translation efficiency, with longer tails promoting more robust translation initiation (Alberts et al., 2015) (Figure 6.2).

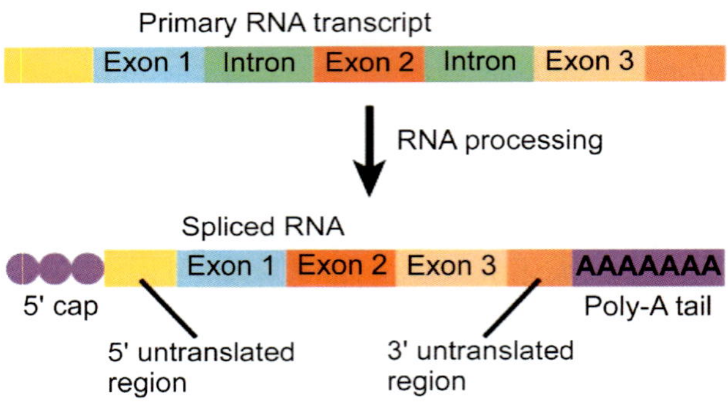

Figure 6.2. *mRNA processing and splicing.*

Source: https://courses.lumenlearning.net/biology/chapter/chapter-9-molecular-biology/.

In addition to these primary processing events, mRNAs can undergo additional modifications to improve their function and stability. One such modification is RNA editing, which alters specific nucleotides within the RNA sequence after transcription. The adenosine deaminase acting on RNA (ADAR) family of enzymes mediates the most common type of RNA editing in eukaryotes: adenosine-to-inosine (A-to-I) editing. The translational machinery recognizes inosine as guanosine, which may result in changes to the amino acid sequence of the encoded protein. RNA editing can also influence splicing, mRNA stability, and localization, contributing to the regulation of gene expression in response to developmental cues and environmental changes (Baynes and Dominiczak, 2009; Licatalosi and Darnell, 2010).

Another important RNA modification is the methylation of adenosine at the N6 position (m6A), which is the most common internal modification in eukaryotic mRNAs. A methyltransferase complex catalyzes m6A methylation, and demethylases can remove it dynamically, making it a reversible modification. This modification has an impact on many aspects of RNA metabolism, such as splicing, translation, and degradation. Specific binding proteins, known as "readers," recognize m6A marks and mediate the downstream effects of the modification. The reversibility of m6A methylation allows cells to rapidly respond to changes in their environment, modulating gene expression patterns accordingly (Licatalosi and Darnell, 2010; Pollard et al., 2022).

Non-coding RNAs, such as ribosomal RNA (rRNA), transfer RNA (tRNA), and small nuclear RNA (snRNA), are also processed as RNA. For example, rRNA processing entails cleaving and chemically modifying a large precursor RNA to produce smaller, functional rRNA molecules that are then assembled into ribosomes. These modifications include methylation and pseudouridylation, both of which are necessary for the ribosome's structural integrity and functionality. Similarly, tRNAs go through extensive processing, such as the removal of 5' and 3' leader sequences, intron splicing, and the addition of a 3' CCA tail, which is required for amino acid attachment during translation. These processing events are required for tRNAs to function correctly during protein synthesis (Alberts et al., 2015).

6.3. REGULATION OF GENE EXPRESSION

Gene expression regulation is a complex and fine-tuned process that is essential for cell function, development, and adaptation. The process involves multiple layers of control, from DNA transcription to protein post-translational modification, to ensure gene expression at appropriate levels, times, and locations within the cell (Ramírez-Clavijo and Montoya-Ortíz, 2013).

Transcriptional control, which determines whether or not a gene is transcribed into RNA, is central to gene expression regulation. This control is achieved through the interaction of transcription factors with specific DNA sequences, such as promoters, enhancers, and silencers. Promoters are DNA regions located upstream of the gene's coding sequence that serve as binding sites for RNA polymerase, the enzyme that synthesizes RNA. RNA polymerase recruitment to the promoter is frequently facilitated by general transcription factors, which assist in correctly positioning the polymerase and initiating transcription (Pollard et al., 2022).

Beyond the core promoter, regions such as enhancers and silencers influence transcription efficiency and rate. Enhancers are DNA sequences that, when bound by specific transcription factors, increase the likelihood of transcription initiation by either facilitating transcriptional machinery assembly or stabilizing the interaction between RNA polymerase and the promoter. These enhancers can act over long distances, sometimes far from the genes they regulate, and can function in a tissue-specific manner, helping to control gene expression spatially and temporally. In contrast, silencers are DNA elements that, when bound by repressor proteins, inhibit transcription by blocking the binding of activators or recruiting co-repressor complexes that modify chromatin to a repressive state (Alberts et al., 2015).

The structure of chromosomes is critical for gene regulation. Eukaryotic cells store DNA in chromatin, a complex of DNA and histone proteins that condenses to form nucleosomes. The level of chromatin compaction influences gene accessibility to the transcriptional machinery. Euchromatin, which is less condensed, is typically associated with active gene expression, whereas heterochromatin, which is more condensed, is associated with gene repression. Epigenetic modifications, such as DNA methylation and histone modifications, are important regulators of chromatin structure. DNA methylation, which usually occurs at CpG dinucleotides, is generally associated with gene silencing because it can prevent the binding of transcription factors or recruit proteins that promote a closed chromatin state (Ramírez-Clavijo and Montoya-Ortíz, 2013).

Histone modifications such as methylation, acetylation, phosphorylation, and ubiquitination affect chromatin dynamics and thus gene expression. For example, histone acetylation by histone acetyltransferases (HATs) generally results in an open chromatin conformation and active transcription. This is because acetylation neutralizes histones' positive charge, reducing their affinity for negatively charged DNA and giving the transcriptional machinery better access to the DNA. In contrast, histone deacetylases (HDACs) remove these acetyl groups, causing chromatin condensation and transcriptional repression. Similarly, histone methylation can activate or repress gene expression, depending on the specific residues modified and the extent of methylation (Licatalosi and Darnell, 2010; Alberts et al., 2015). Post-transcriptional regulation provides another level of control over gene expression. When a gene is transcribed, the resulting RNA goes through several processing steps, including splicing, 5' capping, and 3' polyadenylation. Alternative splicing, which involves joining different exons of a pre-mRNA in various combinations, enables a single gene to produce multiple mRNA isoforms, each of which encodes a different protein variant. Splicing factors tightly regulate this process, which is also influenced by cellular signals and developmental cues. The 5' cap and poly-A tail of mRNAs protect them from degradation while also facilitating translation. MicroRNAs (miRNAs) and RNA-binding proteins can bind to specific sequences in mRNAs and either promote degradation or inhibit translation, further regulating their stability and translation efficiency (Pollard et al., 2022). Gene expression is controlled at the translational level by mechanisms that govern protein synthesis initiation and elongation. The availability of translation initiation factors, the presence of upstream open reading frames (uORFs) in the mRNA, and the mRNA's secondary structure can all affect translation initiation efficiency. Additionally, in response to stress or nutrient deprivation, cells can globally reduce translation by phosphorylating the translation initiation factor eIF2α, thereby conserving resources and altering the proteome to adapt to changing conditions (Baynes and Dominiczak, 2009; Pollard et al., 2022).

Post-translational modifications (PTMs) are a final regulatory checkpoint that fine-tunes protein function and activity. These modifications include phosphorylation, ubiquitination, acetylation, and glycosylation, among others. Kinases mediate phosphorylation, which is a reversible modification that can activate or deactivate proteins, change their localization, or promote their degradation. Ubiquitination, the binding of ubiquitin molecules to proteins, frequently targets the protein for degradation by the proteasome, thereby controlling protein levels within the cell. Other PTMs can influence protein interactions, stability, and function, ensuring that proteins are active only when and where they are required (Ramírez-Clavijo and Montoya-Ortíz, 2013).

RNA localization and decay also control gene expression. Certain mRNAs are selectively transported to specific areas of the cell, where their localized translation enables spatial control of protein production. For example, during development, the localization of mRNAs to specific regions of the embryo is critical for establishing body axes and patterning. Furthermore, mRNA stability is tightly controlled by both cis-elements within the RNA and trans-acting factors, which work together to determine the mRNA's half-life and thus its availability for translation. Degradation pathways, such as the nonsense-mediated decay (NMD) pathway, ensure that defective mRNAs, which could produce abnormal proteins, are rapidly degraded (Alberts et al., 2015). The combination of these different regulatory mechanisms enables cells to precisely control gene expression in response to internal and external signals. This regulation is critical for cell differentiation, development, and adaptation to environmental changes. Dysregulation of gene expression is associated with numerous diseases, including cancer, where mutations in genes encoding transcription factors, chromatin modifiers, or components of the RNA processing machinery can lead to aberrant gene expression profiles (Pollard et al., 2022).

6.4. EPIGENETICS AND CHROMATIN REMODELING

Epigenetics and chromatin remodeling are critical for understanding how gene expression is regulated without affecting the underlying DNA sequence. These processes include DNA and histone protein modifications, as well as dynamic chromatin restructuring, which influence gene accessibility to the transcriptional machinery. Epigenetics research reveals how cells maintain stable gene expression patterns across cell divisions, as well as how environmental factors can cause heritable changes in gene function (Becker and Workman, 2013).

Epigenetics refers to changes such as DNA methylation and histone modifications that help to establish and maintain gene expression states. DNA methylation typically occurs at the 5' position of cytosine residues in CpG dinucleotides, yielding 5-methylcytosine. This modification is catalyzed by DNA methyltransferases (DNMTs), which play critical roles in development, cellular differentiation, and the suppression of transposable elements. The methylation of CpG islands, which are frequently found near gene promoters, is associated with transcriptional repression. The methylated DNA can recruit methyl-binding domain proteins (MBDs), which attract histone deacetylases (HDACs) and other chromatin remodeling factors that induce a closed chromatin conformation, thus silencing gene expression (George et al., 2009).

Histone modifications add an additional layer of epigenetic regulation. Histones, the core proteins that wrap around DNA to form nucleosomes, undergo a variety of post-translational modifications such as methylation, acetylation, phosphorylation, and ubiquitination. These modifications take place primarily on the histone tails that protrude from the nucleosome and can either activate or repress transcription depending on the specific residues and type of modification. Histone acetyltransferases (HATs) catalyze histone acetylation, which is commonly associated with gene activation. Acetylation reduces the positive charge on histones, lowering their affinity for negatively charged DNA and resulting in a more open chromatin structure conducive to transcription. Conversely, deacetylation by HDACs leads to chromatin condensation and transcriptional repression (Ha, 2013; Pollard et al., 2022).

Histone methylation, which is catalyzed by histone methyltransferases (HMTs), is another important modification that can have varying effects on gene expression depending on the circumstances. For example, histone H3 methylation on lysine 4 (H3K4me) is commonly associated with active transcription, especially when this mark is found in promoters or enhancers. Histone H3 methylation at lysine 9 (H3K9me) or lysine 27 (H3K27me) is commonly associated with transcriptional repression. These repressive marks are recognized by specific chromatin-associated proteins, which facilitate the formation of heterochromatin, a densely packed form of chromatin that is normally transcriptionally inactive (Becker and Workman, 2013) (Figure 6.3).

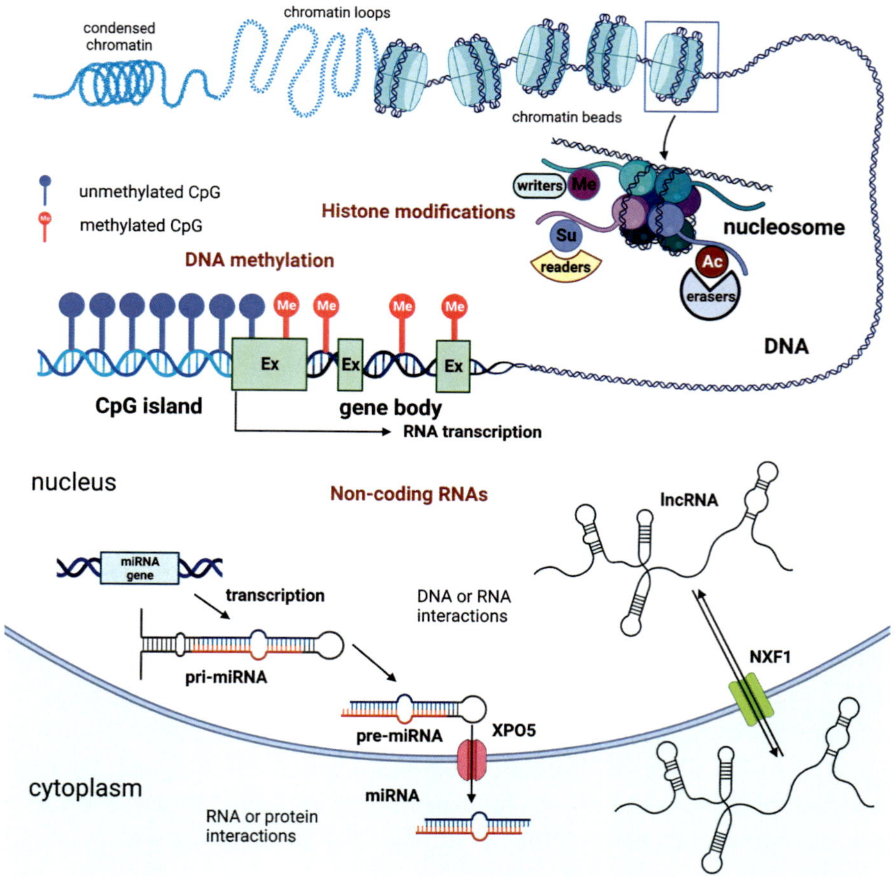

Figure 6.3. *DNA methylation, histone modifications and chromatin remodeling as regulatory mechanisms of epigenetic gene regulation.*

Source: Lossi L, Castagna C, Merighi A. An overview of the epigenetic modifications in the brain under normal and pathological conditions. International Journal of Molecular Sciences. 2024; 25(7):3881. https://doi.org/10.3390/ijms25073881.

Chromatin remodeling is the dynamic alteration of chromatin architecture that affects the accessibility of DNA to transcription factors and regulatory proteins. This process is necessary for proper gene expression regulation, as well as DNA replication, repair, and recombination. Chromatin remodeling complexes, such as the SWI/SNF, ISWI, and CHD families, use energy from ATP hydrolysis to reposition, eject, or restructure nucleosomes. By changing the position of nucleosomes, these complexes can either expose or occlude regulatory regions of DNA, facilitating or hindering the binding of transcription factors and the transcriptional machinery (Becker and Workman, 2013; Alberts et al., 2015).

The interaction of epigenetic modifications and chromatin remodeling is critical for establishing distinct transcriptional programs in various cell types and responding to environmental cues. For example, during cellular differentiation, specific sets of genes are activated or deactivated via coordinated changes in histone modifications and nucleosome positioning. These changes are frequently stable through cell divisions, allowing differentiated cells to retain their identity. Furthermore, in response to external signals, such as stress or nutrient availability, chromatin remodeling complexes can rapidly alter chromatin structure to allow the expression of genes that are critical for adaptation and survival (Ha, 2013).

Epigenetic modifications are also reversible, which is critical for cellular plasticity and adaptability to changing environments. Enzymes like DNA demethylases and histone demethylases can remove methyl groups from DNA and histones, respectively, reversing the repressive marks and possibly reactivating gene expression. Similarly, histone deacetylases can remove acetyl groups, while histone acetyltransferases can reacetylate histones, allowing for dynamic changes between active and repressive chromatin states (George et al., 2009).

Importantly, epigenetic dysregulation and aberrant chromatin remodeling are linked to a variety of diseases, including cancer, neurological disorders, and developmental syndromes. In cancer, for example, tumor suppressor genes can be silenced by hypermethylation of CpG islands, or oncogenes can be misregulated by aberrant histone modifications, driving tumorigenesis. Understanding the mechanisms behind these changes has led to the development of epigenetic therapies, such as DNMT inhibitors and HDAC inhibitors, which aim to restore normal gene expression patterns in cancer cells (George et al., 2009).

Epigenetics and chromatin remodeling research is still revealing the complexities of these regulatory networks, as well as their effects on gene expression, development, and disease. These studies highlight the epigenome's dynamic nature and ability to integrate genetic and environmental signals, resulting in diverse phenotypic outcomes (Alberts et al., 2015).

6.5. POST-TRANSLATION MODIFICATION

Post-translational modifications (PTMs) are critical mechanisms in cell biochemistry that influence protein function, interactions, localization, and stability with other cellular components. These alterations occur after a protein has been synthesized, altering its chemical composition and, as a result, its activity within the cell. PTMs control a wide range of biological activities, including apoptosis, cell cycle progression, signal transduction, and protein degradation.

One of the most prevalent PTMs is phosphorylation, which involves the addition of a phosphate group to certain amino acid residues, usually serine, threonine, or tyrosine. This modification can change the protein's shape, affecting its activity and interactions with other molecules. Protein kinases, for example, phosphorylate target proteins in signal transduction pathways, which causes downstream signaling cascades to be activated or inhibited. The dynamic nature of phosphorylation-dephosphorylation cycles enables cells to adapt quickly to external stimuli, making phosphorylation an important regulatory mechanism in cellular processes (Baynes and Dominiczak, 2009).

Another significant PTM is ubiquitination, which occurs when ubiquitin, a tiny regulatory protein, covalently attaches to lysine residues on target proteins. Ubiquitination often targets proteins for degradation by the proteasome, a large protein complex that degrades unwanted or damaged proteins. However, ubiquitination is not only used to degrade proteins; it can also affect protein activity, localization, and interactions. Monoubiquitination, for example, can influence protein trafficking and DNA repair, but polyubiquitination frequently signals proteasomal destruction, demonstrating this modification's plasticity in cellular regulation (Licatalosi and Darnell, 2010).

Proteins can also acquire carbohydrate moieties through another PTM called glycosylation. Proper protein folding, stability, and cell surface expression depend on this alteration, which mostly takes place in the endoplasmic reticulum and Golgi apparatus. The immune system's ability to recognize pathogens and communicate with other cells can also be impacted by glycosylation patterns. N-linked glycosylation, in particular, is necessary for many secreted and membrane proteins to fold correctly and to maintain their quality control. It occurs when carbohydrates are bonded to asparagine residues (Alberts et al., 2015). Acetylation, the addition of an acetyl group to lysine residues, is a PTM that has a significant impact on protein function, notably in terms of chromatin structure and gene expression. Histone acetylation, for example, neutralizes the positive charge on lysine residues, resulting in a more flexible chromatin shape and greater accessibility of transcriptional machinery to DNA. This alteration is dynamically controlled by histone acetyltransferases (HATs) and histone deacetylases (HDACs), which add and remove acetyl groups. The balance of acetylation and deacetylation is important for gene expression control and has been linked to a variety of illnesses, including cancer (Licatalosi and Darnell, 2010)).

Protein function and gene expression can also be impacted by methylation, which is the addition of methyl groups to arginine or lysine residues. For instance, the exact amino acid residue that is methylated and the quantity of methyl groups added determine whether methylation of histones results in transcription activation or repression. As an illustration, trimethylation of histone H3 at lysine 4 (H3K4me3) is connected to active transcription, while trimethylation at lysine 27 (H3K27me3) is connected to the repression of gene expression. As methyltransferases and demethylases, respectively, add and remove methyl groups, this modification is strictly regulated, guaranteeing exact control over patterns of gene expression (Baynes and Dominiczak, 2009)

Sumoylation, or the attachment of small ubiquitin-like modifier (SUMO) proteins to lysine residues, is a PTM that affects a variety of cellular functions, including nuclear transport, transcriptional control, and DNA repair. Sumoylation, unlike ubiquitination, does not often target proteins for breakdown but rather modifies their activity, stability, or subcellular localization. Sumoylation can also affect protein-protein interactions, acting as a scaffold for the formation of multiprotein complexes. Sumoylation is reversible and controlled by specialized enzymes, allowing cells to fine-tune their responses to a variety of physiological and stress-related signals (Licatalosi and Darnell, 2010).

Palmitoylation, or the addition of palmitic acid to cysteine residues, is a lipid modification that alters protein-membrane interaction and trafficking. This alteration is critical for the appropriate location and function of numerous signaling proteins, including those involved in the Ras and Wnt signaling pathways. Palmitoylation is unlike other lipid modifications in that it is reversible, allowing proteins to cycle between membrane-bound and cytosolic forms, controlling their activity and interactions within the cell (Baynes and Dominiczak, 2009). These examples demonstrate the range and intricacy of PTMs in regulating biological functions. PTMs can function independently or in combination, resulting in a highly dynamic and sophisticated network of regulatory processes. PTMs' ability to rapidly change protein function and localization in response to cellular signals highlights their importance in maintaining cellular homeostasis and responding to environmental changes. As research in this subject progresses, more PTMs and their functions in cellular regulation are likely to be uncovered, adding to our understanding of the molecular mechanisms that control cell biology.

6.6. MICRORNAS AND GENE SILENCING

MicroRNAs (miRNAs) are small, non-coding RNA molecules that have received a lot of interest in the field of cell biochemistry because of their important function in gene regulation. These molecules, which are generally 20-24 nucleotides long, act as strong modulators of gene expression by targeting messenger RNA (mRNA) for destruction or translational suppression, a process known as gene silence. This complicated method demonstrates the substantial impact miRNAs have on cellular functions such as differentiation, proliferation, metabolism, and aptopsis.

MiRNA biogenesis begins in the nucleus, where they are transcribed by RNA polymerase II or III to become primary miRNAs. The Drosha-DGCR8 complex then processes these pri-miRNAs into precursor miRNAs (pre-miRNAs), which are around 70 nucleotides long. After nuclear processing, pre-miRNAs are exported to the cytoplasm, where they are further cleaved by the enzyme Dicer to form mature miRNAs. The mature miRNA is then integrated into the RNA-induced silencing complex (RISC), where it directs the complex to complementary sequences in target mRNAs, causing gene silencing via mRNA destruction or translational repression (Bartel, 2004) (Figure 6.4).

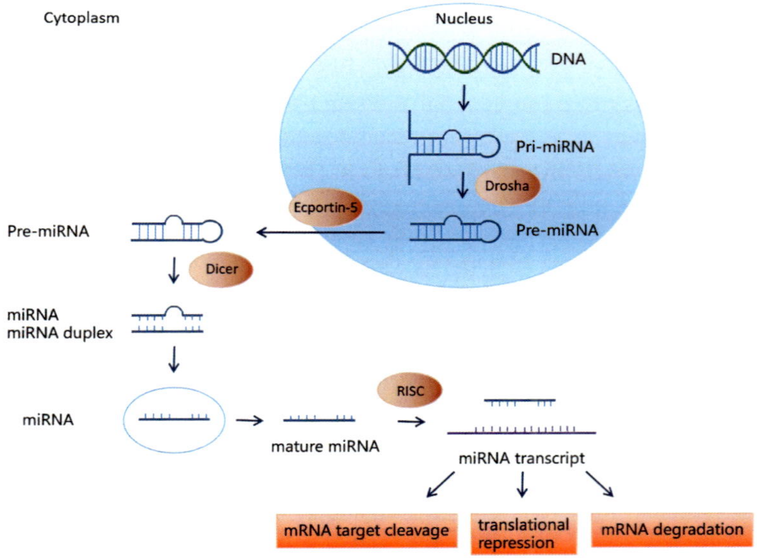

Figure 6.4. *miRNA biogenesis and related gene silencing mechanisms.*

Source: Liu J, Zhou F, Guan Y, Meng F, Zhao Z, Su Q, Bao W, Wang X, Zhao J, Huo Z, et al. (2022). The biogenesis of miRNAs and their role in the development of amyotrophic lateral sclerosis. Cells, 11(3):572. https://doi.org/10.3390/cells11030572.

One of the most important elements of miRNA-mediated gene silencing is its selectivity and adaptability. MiRNAs often bind to the 3' untranslated region (3' UTR) of target mRNAs, where they can exert their effects based on the degree of complementarity between the miRNA and mRNA. Perfect or near-perfect complementarity typically results in mRNA cleavage and destruction, but partial complementarity frequently causes translational repression without mRNA degradation (Licatalosi and Darnell, 2010)). This dual approach enables miRNAs to fine-tune protein expression levels within the cell, thereby contributing to the exact regulation of diverse metabolic pathways.

MiRNAs have an important function in cell biochemistry, including crucial physiological and pathological processes. For example, in the context of cellular differentiation, certain miRNAs have been identified as key regulators of cell destiny. The let-7 miRNA family, for example, is known to promote differentiation by inhibiting the expression of genes that maintain stem cell pluripotency (Roush and Slack, 2008). Similarly, miR-17-92 cluster members have been linked to regulating the balance of cell proliferation and apoptosis, which influences carcinogenesis (Pollard et al., 2024).

In addition to their role in normal cellular functions, miRNAs have been extensively researched for their role in a variety of diseases, particularly cancer. Dysregulation of miRNAs is frequent in malignant cells, where they can function as oncogenes or tumor suppressors depending on the situation. For example, miR-21 is often increased in several malignancies and has been shown to promote tumor growth by targeting and reducing the expression of tumor suppressor genes including PTEN and PDCD4 (Chan et al., 2005). MiRNAs, such as miR-34a, which is frequently downregulated in cancer, act as tumor suppressors by causing cell cycle arrest and apoptosis by inhibiting oncogenes such as MYC and BCL2 (Gagneux et al., 2022).

The ramifications of miRNA-mediated gene silencing go beyond oncology. In neurobiology, miRNAs are thought to regulate synaptic plasticity and neurogenesis. MiR-132, for example, has been demonstrated to stimulate neurite outgrowth by targeting the GTPase-activating protein p250GAP, which promotes synaptic function (Punekar, 2018). Furthermore, miRNAs are increasingly becoming recognized as important players in metabolic control. MiR-33, which co-transcribes with the sterol regulatory element-binding protein (SREBP) genes, controls cholesterol homeostasis and fatty acid metabolism by targeting genes involved in cholesterol efflux and fatty acid oxidation (Alberts et al., 2015).

The biochemical processes by which miRNAs silence genes are also quite interesting. MiRNAs and their target mRNAs normally interact by base-pairing the miRNA's seed region (nucleotides 2-7) with complementary sequences inside the mRNA's 3' UTR. This relationship is aided by the RISC, which contains Argonaute proteins as its key components. Argonaute proteins serve an important role in miRNA targeting and modulating the following gene-silencing effects. Structural investigations revealed that Argonaute proteins provide a binding pocket for the 5' end of the miRNA, positioning the seed region for optimum interaction with the target mRNA (Ramírez-Clavijo and Montoya-Ortíz, 2013). The specificity of miRNA-target interactions depends on this exact placement, which enables miRNAs to control several mRNAs at once and hence coordinate intricate gene networks inside the cell.

The fact that miRNAs can target numerous mRNAs and that several miRNAs can target a single mRNA adds to the complexity of miRNA-mediated gene silence. Because of these multiple connections, miRNAs can operate as nodes in gene regulatory networks, combining signals from several pathways and regulating the expression of many genes simultaneously. MiR-155, for instance, targets a broad range of mRNAs implicated in inflammation, immune cell differentiation, and cytokine production, and has been identified as a fundamental regulator of immunological responses (Gomperts et al., 2009). The significance of miRNAs in preserving cellular homeostasis and reacting to external stimuli is highlighted by this network-based regulation.

The potential of miRNAs as therapeutic targets has also been brought to light by research. Modulating miRNA levels in cells is a viable approach to treating diseases linked to aberrant gene expression because of their pivotal function in regulating gene expression. For example, preclinical models of cardiovascular disease and cancer have demonstrated potential for antisense oligonucleotides (ASOs) called antagomirs, which are targeted to suppress particular miRNAs (Becker and Workman, 2013). Furthermore, in disorders where certain miRNAs are downregulated, the administration of synthetic miRNAs or miRNA mimics has been investigated as a way to restore normal gene expression.

Consequently, miRNAs play a crucial role in the gene regulation system of cell biochemistry. miRNAs affect a variety of cellular processes, including differentiation, proliferation, apoptosis, and metabolism, by modulating gene expression through mRNA degradation or translational repression. Their importance in both health and sickness is highlighted by the numerous disorders that are linked to their dysregulation. Research on the processes behind miRNA-mediated gene silencing and their wider effects on cellular function is still very active, with significant applications in biology and medicine.

6.7. GENE EXPRESSION IN DEVELOPMENT AND DISEASE

In cell biochemistry, gene expression is a basic process that controls the complex dynamics of development and illness. The genetic information is transcribed from DNA into messenger RNA (mRNA) and then translated into useful proteins through this process. For cells to function normally, gene expression must be precisely regulated. Any deviations from this balance can cause diseases such as cancer, neurological disorders, and metabolic syndromes, as well as developmental abnormalities.

Gene expression is strictly regulated during development to guarantee that cells differentiate and take on specific roles. In this case, the intricate interactions of transcription factors, epigenetic changes, and non-coding RNAs regulate the expression of genes. Proteins known as transcription factors attach to particular DNA sequences to either promote or prevent the transcription of a given gene. For instance, the transcription factor MyoD plays a critical role in the differentiation of muscle cells by initiating a series of genes specific to muscles that are required for the growth of muscular tissue (Jin et al., 2016). Similar to this, the transcription factor Oct4 is necessary to keep embryonic stem cells pluripotent, which enables them to develop into numerous kinds of cells (Becker and Workman, 2013).

Histone alterations and other epigenetic changes, such as DNA methylation, are also essential for controlling the expression of certain genes during development. Due to these changes to the chromatin structure, the transcriptional machinery may access different parts of the genome to varying degrees. DNA methylation usually inhibits transcription factors from attaching to the DNA, which reduces the expression of the gene. On the other hand, because histone acetylation loosens the chromatin structure and makes it easier for transcription factors to reach the DNA, it usually improves gene expression. Particular processes, such as genomic imprinting, in which particular genes are expressed in a parent-of-origin-specific way, and X-chromosome inactivation in female mammals depend on these epigenetic mechanisms (Pollard et al., 2022)).

In the setting of disease, abnormal gene expression can cause pathological conditions. Cancer, for example, is frequently characterized by dysregulation of gene expression, with oncogenes overexpressed and tumor suppressor genes suppressed. Overexpression of oncogenes like MYC promotes uncontrolled cell proliferation and survival, which contributes to carcinogenesis (George et al., 2009). In contrast, suppressing tumor suppressor genes such as p53 eliminates key constraints on cell proliferation and death, increasing the malignant phenotype (George et al., 2009). Cancer cells typically exhibit epigenetic changes, such as aberrant DNA methylation patterns and histone modifications, which result in the improper expression or suppression of genes that promote tumor development and spread.

Neurodegenerative illnesses, such as Alzheimer's and Parkinson's, are also associated with gene expression alterations. Alzheimer's disease alters the expression of genes involved in amyloid precursor protein (APP) processing and tau protein phosphorylation, resulting in the buildup of amyloid plaques and neurofibrillary tangles, which are disease hallmarks. Mutations in the gene that encodes alpha-synuclein, a protein that clumps in the brains of Parkinson's patients, cause overexpression and the creation of toxic aggregates, which contribute to neuronal death (Ha, 2013; Pollard et al., 2022). These examples show how dysregulated gene expression leads to the etiology of neurodegenerative diseases.

Metabolic disorders, such as diabetes and obesity, are also associated with altered gene expression. Type 2 diabetes alters the expression of genes involved in insulin signaling and glucose metabolism, resulting in insulin resistance and decreased glucose uptake by cells. FOXO1, for example, regulates gluconeogenesis and insulin sensitivity. FOXO1 is inappropriately activated in diabetic patients, resulting in increased glucose synthesis by the liver and worsening hyperglycemia (Becker and Workman, 2013). Obesity causes dysregulation of genes involved in lipid metabolism and adipogenesis, which contributes to excess fat accumulation and the development of obesity-related problems (Alberts et al., 2015).

Furthermore, environmental influences can influence gene expression by altering the epigenome and the transcriptional landscape of cells. For example, exposure to environmental pollutants such as tobacco smoke or heavy metals can cause epigenetic modifications that disrupt normal gene expression patterns, raising the risk of developing cancer and cardiovascular disease (Alberts et al., 2015). These findings show the complicated interplay of gene expression, environmental variables, and disease risk.

Chapter 7

Protein Structure and Function

Contents

Protein structure and function are fundamental to cellular biochemistry, supporting almost all biological processes. Proteins are composed of amino acids linked together by peptide bonds, resulting in polypeptide chains that fold into specific three-dimensional structures required for their function. Proteins are organized into four hierarchical levels: primary, secondary, tertiary, and quaternary (Alberts et al., 2015).

The primary structure is the linear sequence of amino acids that determines the subsequent folding patterns. Polypeptide chains form secondary structures like α-helices and β-sheets through hydrogen bonds between their backbone atoms. The tertiary structure refers to the polypeptide's overall three-dimensional folding, which is stabilized by a variety of interactions such as hydrophobic interactions, disulfide bridges, and ionic bonds. Some proteins form quaternary structures, where multiple polypeptide subunits assemble into functional complexes (Alberts et al., 2015).

A protein's function depends on its specific folding and conformation. Enzymes, for example, are a type of protein that catalyzes biochemical reactions with high specificity based on the shapes of their active sites. Structural proteins, such as collagen, provide mechanical support, whereas transport proteins, such as hemoglobin, help molecules move around the body. Misfolding of proteins can cause diseases, demonstrating the complex relationship between protein structure and function. Understanding protein structure is thus critical for deciphering cellular mechanisms and devising therapeutic strategies.

7.1. PROTEIN FOLDING AND STABILITY

Protein folding is a complex and tightly regulated process that begins when the nascent polypeptide chain emerges from the ribosome. This intricate process is required for protein functional maturation because the protein's biological activity is determined by its three-dimensional structure. The folding process is affected by the primary sequence of amino acids, the cellular environment, and various molecular chaperones and enzymes that facilitate proper folding and prevent misfolding (Baynes and Dominiczak, 2009).

As the ribosome synthesizes the polypeptide chain, it begins to fold back into its native conformation. Anfinsen's dogma states that the amino acid sequence in a polypeptide chain contains all of the information required for the protein to achieve its final structure. This sequence specifies the folding pathway, which is influenced by non-covalent interactions such as hydrogen bonds, ionic interactions, van der Waals forces, and hydrophobic effects. These interactions lead to the formation of secondary structures like α-helices and β-sheets, which are the first steps in folding. These elements further organize into supersecondary structures, and eventually into the tertiary structure, where the entire polypeptide chain adopts its functional form (Alberts et al., 2015; Balchin et al., 2016).

However, protein folding in the cellular environment is not a simple or spontaneous process. The crowded and complex environment of the cell, also known as the cellular "crowding effect," can impede proper folding and increase the likelihood of misfolding or aggregation. To ensure proper folding, cells use a complex network of molecular chaperones and folding enzymes. Chaperones, like the heat shock proteins (Hsp70 and Hsp90 families), bind to nascent or partially folded polypeptides and stabilize them in a folding-competent state. These chaperones prevent the aggregation of hydrophobic regions exposed during the early stages of folding, thus facilitating the correct folding pathway (Hartl et al., 2011; Pollard et al., 2022).

The timing and regulation of protein folding events is an important aspect of the process. Co-translational folding, which occurs during polypeptide synthesis, allows specific domains of the protein to fold independently before the entire chain is completed. This stepwise folding reduces the possibility of misfolding by limiting the number of folding intermediates present at any given time. Ribosome-associated chaperones, such as the nascent polypeptide-associated complex (NAC) and the ribosome-associated complex (RAC), play critical roles in this process by interacting with the emerging polypeptide chain and guiding it along the desired folding pathway (Baynes and Dominiczak, 2009).

Post-translational modifications (PTMs) also influence protein folding and stability. PTMs, such as glycosylation, phosphorylation, and disulfide bond formation, can affect folding by altering the chemical properties of the polypeptide chain. Disulfide bonds, for example, formed by the oxidation of cysteine residues, help to stabilize protein tertiary and quaternary structures by covalently linking different regions of the polypeptide chain. Enzymes, such as protein disulfide isomerase (PDI) catalyze the formation and rearrangement of these bonds, ensuring that the proper disulfide linkages are formed during the folding process (Balchin et al., 2016).

Protein stability, defined as a protein's ability to retain its native conformation under physiological conditions, is inextricably linked to the folding process. The native state of a protein is a thermodynamically stable conformation that corresponds to the lowest free energy state. Proteins, on the other hand, are dynamic molecules with frequent minor structural variations. A protein's stability is influenced by several factors, including its primary sequence, the presence of stabilizing interactions (e.g., hydrogen and disulfide bonds) and the cellular environment (Alberts et al., 2015; Balchin et al., 2016).

Proteins can be destabilized by a variety of factors, including temperature changes, pH changes, and the presence of denaturants, resulting in unfolding or misfolding. Misfolded proteins are frequently identified and targeted for degradation by the cell's quality control systems, including the ubiquitin-proteasome system. However, if misfolded proteins escape degradation, they can aggregate and form insoluble fibrils, which are associated with various neurodegenerative diseases, including Alzheimer's and Parkinson's diseases (Pollard et al., 2022).

Molecular chaperones not only help with protein folding, but they also play an important role in maintaining protein stability over time. They can recognize and refold misfolded proteins, preventing aggregation and facilitating the recovery of functional proteins. Furthermore, certain chaperones participate in the assembly of multi-subunit complexes, ensuring that proteins achieve their correct quaternary structure, which is essential for their stability and function (Baynes and Dominiczak, 2009; Pollard et al., 2022).

7.2. PROTEIN DOMAINS AND MOTIFS

Protein domains and motifs are fundamental as they underpin the structural and functional diversity of proteins. Understanding these elements is critical for understanding how proteins perform a wide range of biological functions, from catalysis and signaling to structural support and transport. Protein domains are distinct structural units within a protein, frequently associated with specific functions, whereas motifs are short, conserved amino acid sequences that contribute to the protein's overall structure or function. They provide a framework for the modular architecture of proteins, which allows for complexity and specialization within cellular environments (Pollard et al., 2022) (Figure 7.1).

Protein domains can be thought of as protein building blocks, representing evolutionary units that can fold independently to form stable three-dimensional structures. These domains are frequently conserved across different proteins and organisms, indicating that they have been preserved throughout evolution due to their critical functional roles. For example, the SH2 domain, a classic example of a protein domain, participates in signal transduction pathways by binding specifically to phosphorylated tyrosine residues on target proteins (Schaeffer et al., 2016). This interaction facilitates the formation of multiprotein complexes required for intracellular signal propagation. Similarly, the immunoglobulin domain of antibodies plays an important role in the immune response by recognizing and binding to specific antigens.

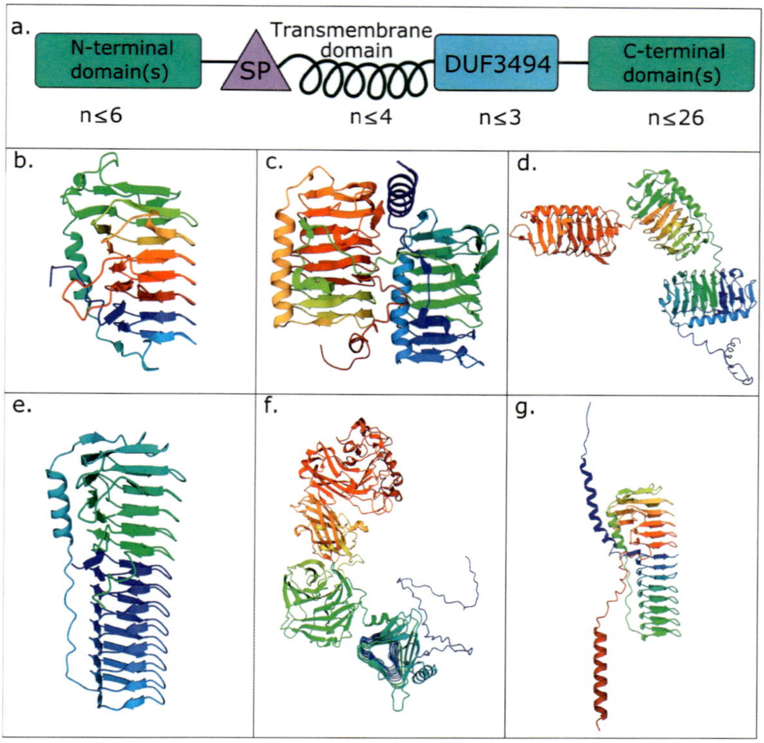

Figure 7.1. *Motifs and domains are parts of full-length proteins with defined structures.*

Source: Winder JC, Boulton W, Salamov A, Eggers SL, Metfies K, Moulton V, Mock T. (2023). Genetic and structural diversity of prokaryotic ice-binding proteins from the central arctic ocean. Genes, 14(2):363. https://doi.org/10.3390/genes14020363.

Domains are more than just structural entities; they also provide specific biochemical properties to proteins. The catalytic domains of enzymes, such as the serine protease domain found in trypsin, chymotrypsin, and other proteases, provide the chemical environment required for catalysis. The conserved serine residue within this domain plays a direct role in peptide bond cleavage, demonstrating how a domain's structure influences its function. Furthermore, domains can serve as scaffolds, arranging other domains within the same protein or mediating interactions with other proteins, nucleic acids, or small molecules (Bhattacharyya et al., 2006). Because of their modular nature, proteins can evolve new functions by recombining existing domains, which is known as domain shuffling or domain recombination. In contrast, motifs are usually shorter than domains and do not fold into stable structures on their own. Instead, motifs are frequently embedded within domains, adding to their structural stability or functional activity. These motifs are highly conserved sequences that aid in the binding of ligands, substrates, and other proteins. For example, the zinc finger motif, which is distinguished by a conserved pattern of cysteine and histidine residues, coordinates a zinc ion to maintain its fold. This motif is commonly found in DNA-binding proteins, where it facilitates specific interactions with DNA and thus regulates gene expression (Baynes and Dominiczak, 2009; Boal, 2012).

Another well-known motif is the helix-turn-helix, which plays an important role in DNA recognition and binding by transcription factors. The motif consists of two α-helices separated by a short turn, with one helix recognizing and binding to the major groove of DNA. The amino acid sequence of the recognition helix determines DNA binding specificity, allowing for precise control over transcriptional regulation (Bhattacharyya et al., 2006; Schaeffer et al., 2016). This demonstrates how even small motifs can have a significant impact on a protein's biological activity by controlling specific molecular interactions.

The interaction between protein domains and motifs is visible in the multidomain architecture of many proteins, where multiple domains and motifs work together to produce complex biological functions. Kinases, for example, are enzymes that transfer phosphate groups from ATP to specific substrates and frequently contain multiple domains, including a catalytic kinase domain and regulatory domains that control kinase activity. Within these domains, motifs such as the ATP-binding motif and the activation loop motif are critical for the proper alignment and phosphorylation of substrates, ensuring precise regulation of cellular signaling pathways (Schaeffer et al., 2016; Pollard et al., 2022).

Protein domains and motifs play important roles in the evolution of protein function. Proteins can acquire new domains via gene duplication and subsequent divergence, resulting in the development of novel functions. The modular nature of signaling proteins, such as those in the MAP kinase pathway, exemplifies this evolutionary plasticity, as domain combinations allow for the integration and processing of multiple signaling inputs, resulting in a coordinated cellular response. Similarly, motifs can evolve to gain new functionalities or to be recognized by different proteins, adding an extra layer of regulation and complexity (Bhattacharyya et al., 2006; Baynes and Dominiczak, 2009).

Understanding the principles that govern the function and evolution of protein domains and motifs is critical for biomedical research, especially in disease models. Many genetic disorders are caused by mutations in specific domains or motifs, which disrupt protein function. Mutations in the DNA-binding domain of p53, a tumor suppressor protein, are one of the most common causes of cancer. These mutations frequently impair the ability of p53 to bind DNA and activate genes involved in cell cycle arrest and apoptosis, resulting in uncontrolled cell proliferation (Boal, 2012). Drug design has shown that targeting specific domains or motifs within proteins is an effective strategy. Inhibitors that mimic the natural ligands or substrates of these domains can selectively inhibit protein activity. For example, kinase inhibitors that target the ATP-binding pocket of the kinase domain were successfully developed for the treatment of various cancers, demonstrating the therapeutic potential of targeting protein domains (Bhattacharyya et al., 2006).

7.3. PROTEIN-LIGAND INTERACTIONS

Protein-ligand interactions are essential for a variety of biological processes, including the function of enzymes, receptors, and many other macromolecules in cellular biochemistry. These interactions are characterized by their specificity and affinity, which are determined by the complementarity of the ligand and the binding site on the protein (Figure 7.2).

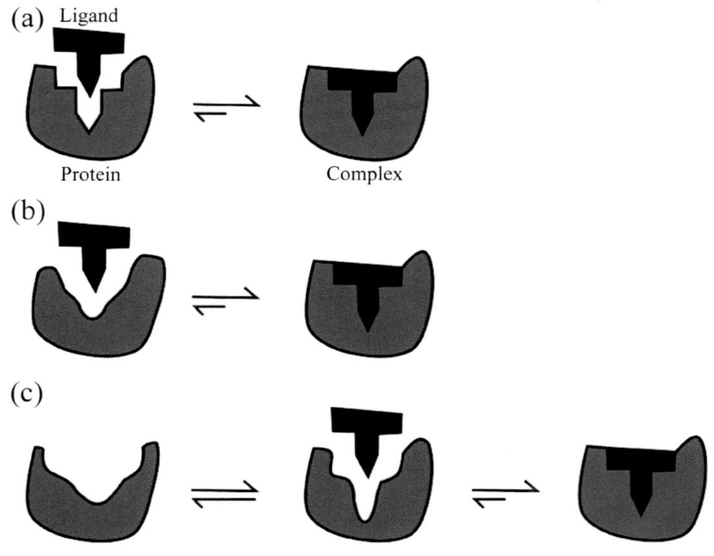

Figure 7.2. *Representation of protein–ligand interactions.*

Source: Du X, Li Y, Xia Y-L, Ai S-M, Liang J, Sang P, Ji X-L, Liu S-Q. (2016). Insights into protein–ligand interactions: Mechanisms, models, and methods. International Journal of Molecular Sciences, 17(2):144. https://doi.org/10.3390/ijms17020144): The figure shows (a) Lock-and-key; (b) Induced fit; and (c) Conformational selection.

Understanding these interactions requires investigating the structural and dynamic aspects of proteins, the nature of ligands, and the thermodynamic and kinetic principles that govern binding events (Du et al., 2016).

Non-covalent interactions such as hydrogen bonds, ionic interactions, van der Waals forces, and hydrophobic effects are primarily responsible for ligand binding to proteins. These interactions occur within a specific region of the protein known as the binding site, which is typically a pocket or groove on its surface. The binding site is shaped and chemically tuned to recognize and interact with the ligand, a process commonly described by the "lock-and-key" or "induced fit" models (Pollard et al., 2022).

Emil Fischer proposed the lock-and-key model in 1894, which suggests that the protein's ligand and binding site have complementary shapes, similar to a key fitting into a lock. This model emphasizes the ligand-protein complementarity that exists before their interaction. However, the induced fit model, proposed later by Daniel Koshland, offers a more dynamic perspective, implying that ligand binding causes conformational changes in the protein, optimizing the fit between the ligand and the binding site. This flexibility is important because it allows the protein to adapt to different ligands or fine-tune its activity in response to binding (Baynes and Dominiczak, 2009).

One of the most well-studied examples of protein-ligand interactions is enzyme-substrate binding, in which the enzyme's active site binds to a specific substrate, catalyzing a chemical reaction. The three-dimensional arrangement of amino acid residues within the active site contributes significantly to the specificity of this interaction. These residues interact with the substrate through a variety of non-covalent interactions, positioning the substrate in a way that reduces the reaction's activation energy, thereby increasing the rate. The efficiency and specificity of enzyme-substrate interactions are critical for cellular metabolism, signaling, and regulation (Salentin et al., 2015).

Understanding protein-ligand interactions is largely dependent on thermodynamics. The dissociation constant (K_d) provides a quantitative description of the binding affinity, or the strength of the interaction, between a protein and its ligand. The higher the protein's affinity for the ligand, the lower the K_d value. This affinity is a reflection of the Gibbs free energy change (ΔG) associated with the binding process, where a negative ΔG indicates a spontaneous and favorable binding interaction. The heat exchange and change in disorder during the binding event are represented by the enthalpic (ΔH) and entropic (ΔS) contributions, which together account for the total free energy change. Favorable entropic contributions usually come from the release of water molecules or other ions from the binding site as the ligand binds, while favorable enthalpic contributions usually arise from the formation of non-covalent bonds

(Salentin et al., 2015; Du et al., 2016). The kinetics of protein-ligand interactions are equally important because they reveal the rate at which these interactions occur. The association rate constant (k_on) and dissociation rate constant (k_off) describe the ligand binding and release rates, respectively. High-affinity interactions are defined by a low K_d, which frequently corresponds to a slow dissociation rate, implying that the ligand remains bound to the protein for a longer time (Du et al., 2016).

Several factors in the cellular environment influence protein-ligand interactions as well. For example, allosteric regulation occurs when a ligand binds to a specific site on a protein, causing conformational changes that influence the binding of another ligand to another site. This type of regulation is common in enzymes and receptors, where binding an allosteric effector can increase or decrease protein activity. Hemoglobin, the oxygen-carrying protein in red blood cells, is a classic example of how oxygen binding at one site influences binding affinity at other sites through cooperative interactions (Salentin et al., 2015).

Water molecules and the protein's hydration shell also play an important role in protein-ligand interactions. Water molecules can mediate interactions between proteins and ligands, often acting as bridges in hydrogen bonding networks. The displacement of water molecules during ligand binding can have a significant effect on the entropic component of binding free energy. In many cases, the reorganization of the hydration shell during ligand binding is an important determinant of binding affinity and specificity (Boal, 2012).

The dynamic nature of protein-ligand interactions makes them even more complex. Proteins are not rigid structures, so conformational changes can affect ligand binding. These fluctuations can either create new binding sites or alter the accessibility of existing ones, influencing binding affinity and specificity. Nuclear magnetic resonance (NMR) spectroscopy and molecular dynamics simulations have provided valuable insights into these dynamic processes, revealing that ligand binding can involve multiple intermediate states and pathways (Alberts et al., 2015; Du et al., 2016).

Furthermore, the cellular context, which includes pH, ionic strength, and the presence of competing ligands or cofactors, influences protein-ligand interactions significantly. In vivo, these interactions take place within the cell's crowded and diverse environment, rather than in isolation. This environment can change the binding affinity and kinetics through various mechanisms, such as changes in protein conformation, the presence of other interacting proteins, or changes in the local concentration of ligands (Du et al., 2016; Pollard et al., 2022).

7.4. ENZYME KINETICS AND INHIBITION

Enzyme kinetics and inhibition provide information about how enzymes catalyze chemical reactions and how their activity can be regulated. Enzymes are biological catalysts that accelerate biochemical reactions by lowering the activation energy required to convert reactants into products. The study of enzyme kinetics entails understanding the rates of these reactions and how they change in response to varying conditions. Inhibition, on the other hand, refers to the decrease in enzyme activity caused by specific molecules. It is critical for regulating metabolic pathways and therapeutic drug design (Guengerich, 2023) (Figure 7.3).

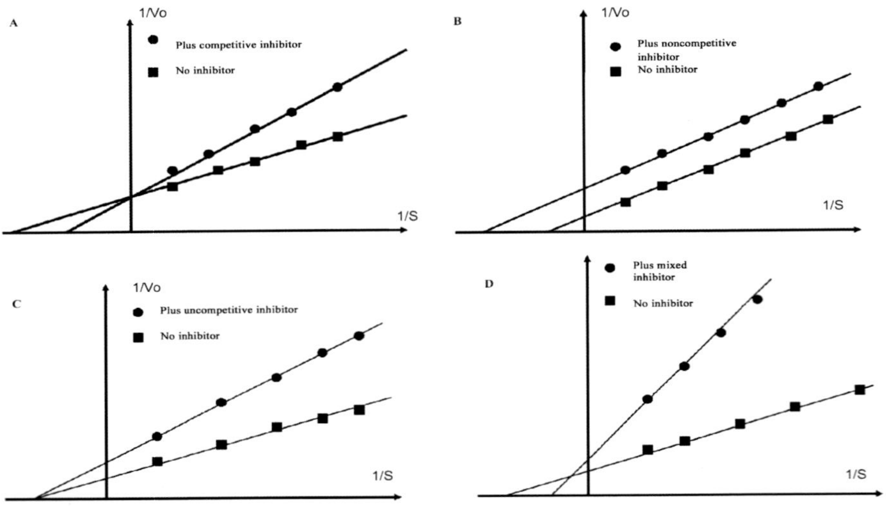

Figure 7.3. *Enzyme kinetics.*

Source: Lopina, O. D. (2017). Enzyme Inhibitors and Activators. InTech. doi: 10.5772/67248.

The Michaelis-Menten model, named after Leonor Michaelis and Maud Menten, who developed it in the early twentieth century, is commonly used to describe the fundamental principles of enzyme kinetics. According to this model, an enzyme (E) binds to a substrate (S) to form an enzyme-substrate complex (ES), which then converts into a product (P) and releases the enzyme. This reaction can be summarized as: $E + S \rightleftharpoons ES \rightarrow E + P$. The reaction velocity (v), or the rate at which the product is formed, is determined by the substrate ([S]) and enzyme concentrations. When the substrate concentration is significantly higher than the enzyme concentration, the enzyme becomes saturated with substrate and the reaction rate reaches its maximum velocity (V_{max}) (Baynes and Dominiczak, 2009).

The Michaelis-Menten equation, $v = V_{max}[S]/K_m + [S]$, describes the relationship between the reaction velocity (v) and the substrate concentration ($[S]$). Here, K_m stands for the Michaelis constant, defined as the substrate concentration at which the reaction velocity is half of Vmax. A lower K_m indicates higher affinity, meaning the enzyme can achieve half of its maximum activity at a lower substrate concentration. It gives an indication of the enzyme's affinity for its substrate. The enzyme's catalytic efficiency at full substrate saturation is indicated by the Vmax, which is directly proportional to the concentration of the enzyme. These parameters are crucial for defining the physiological roles of enzymes in metabolic pathways and for assessing the activity of enzymes (Kornbrust et al., 2012; Pollard et al., 2022).

Enzyme inhibition is an important regulatory mechanism that can occur in several ways, including competitive, non-competitive, uncompetitive, and mixed inhibition. Each mode of inhibition affects enzyme kinetics differently, providing insight into the enzyme's mechanism and function (Guengerich, 2023).

In competitive inhibition, the inhibitor molecule resembles the substrate and competes for binding to the enzyme's active site. Because the substrate and inhibitor cannot bind at the same time, increasing the substrate concentration can counteract the inhibitor's effects. Thus, in competitive inhibition, the Vmax remains constant, but the apparent K_m rises as a higher substrate concentration is required to reach half of V_{max}. Malonate, a structurally similar compound to succinate, the enzyme's natural substrate, inhibits succinate dehydrogenase, which is a classic example of competitive inhibition (Baynes and Dominiczak, 2009).

Uncompetitive inhibition occurs when an inhibitor binds to an allosteric site on the enzyme that differs from the active site, causing a conformational change that reduces the enzyme's catalytic efficiency. Unlike competitive inhibition, non-competitive inhibition affects the Vmax but not the Km because the substrate can still bind to the enzyme. However, the enzyme-substrate-inhibitor complex formed during non-competitive inhibition is less efficient at converting substrate to product. Cyanide inhibits cytochrome c oxidase by binding to a site other than the active site, disrupting the enzyme's activity. This is an example of uncompetitive inhibition (Kornbrust et al., 2012; Guengerich, 2023). Non-competitive inhibition occurs when the inhibitor only binds to the enzyme-substrate complex and not the free enzyme. This binding typically occurs at an allosteric site after the substrate has been bound, stabilizing the enzyme-substrate complex and preventing it from converting to the product. Uncompetitive inhibition reduces both Vmax and K_m by effectively removing the enzyme-substrate complex from the reaction pathway, requiring lower substrate concentrations to achieve half of the reduced Vmax. This type of inhibition is less common but occurs in some enzyme systems, such as the inhibition of alkaline phosphatase by phenylalanine (Pollard et al., 2022).

Mixed inhibition refers to an inhibitor that can bind to both the free enzyme and the enzyme-substrate complex but with different affinities. This type of inhibition influences both K_m and Vmax, depending on the inhibitor's relative affinity for the enzyme and the enzyme-substrate complex. Mixed inhibition is a more general case that can reduce catalytic activity while changing substrate binding, providing a versatile mechanism for enzyme regulation. An example of mixed inhibition is the inhibition of some dehydrogenases by heavy metals, such as mercury, which can bind to different forms of the enzyme (Guengerich, 2023).

Enzyme kinetics and inhibition play critical roles in drug development and therapeutic interventions. Many pharmaceuticals are enzyme inhibitors, which are intended to target specific enzymes involved in disease pathways. Statins, a type of drug used to lower cholesterol, act as competitive inhibitors of HMG-CoA reductase, a key enzyme in cholesterol biosynthesis. Statins reduce the production of cholesterol by inhibiting this enzyme, lowering blood cholesterol levels and decreasing the risk of cardiovascular disease (Baynes and Dominiczak, 2009).

Another example is the use of protease inhibitors to treat viral infections, such as HIV. These inhibitors target the viral protease enzyme, which is responsible for converting viral polyproteins into mature, functional forms needed for viral replication. By binding to the active site of the protease, these inhibitors prevent the maturation of viral proteins, thus blocking virus replication and reducing viral load in patients (Boal, 2012).

Furthermore, the principles of enzyme kinetics are critical for understanding metabolic regulation and enzyme deficiencies in various diseases. Many metabolic disorders are caused by mutations in enzyme-encoding genes, which alter enzyme kinetics and disrupt metabolic pathways. For example, in phenylketonuria (PKU), a mutation in the gene encoding phenylalanine hydroxylase results in an enzyme with a lower affinity for its substrate, phenylalanine. This decreased affinity, reflected by an increased K_m, leads to the accumulation of phenylalanine, causing toxic effects in the brain and severe intellectual disability if not treated (Guengerich, 2023; Kornbrust et al., 2012).

7.5. PROTEIN DEGRADATION PATHWAYS

Protein degradation pathways are key processes in cell biochemistry that ensure normal protein turnover, maintain cellular homeostasis, and regulate diverse cellular functions. These pathways not only aid in the elimination of damaged or misfolded proteins, but they also play an important role in managing the amounts of functioning proteins, which influence cellular signaling, metabolism, and survival.

In eukaryotic cells, there are two principal protein degradation pathways: the ubiquitin-proteasome system (UPS) and the autophagy-lysosome pathway, each of which has distinct processes and biological significance (Figure 7.4).

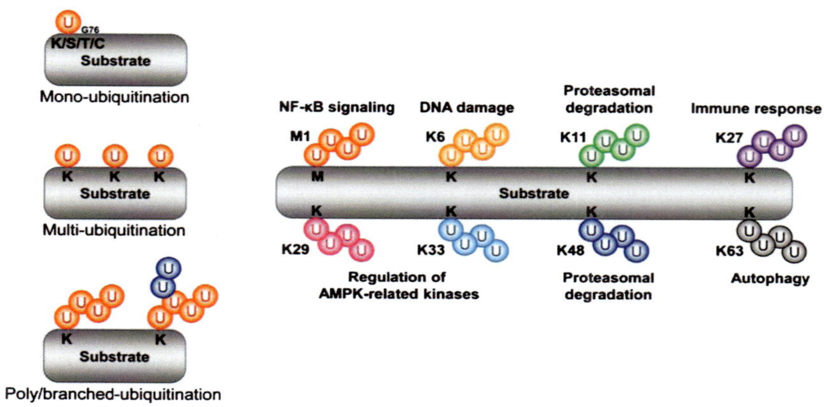

Figure 7.4. *Protein degradation pathways in eukaryotic cells.*

Source: Kim Y, Kim E-K, Chey Y, Song M-J, Jang HH. (2023). Targeted protein degradation: Principles and applications of the proteasome. Cells, 12(14), 1846. https://doi.org/10.3390/cells12141846).

The ubiquitin-proteasome system is responsible for the disintegration of misfolded, short-lived, or regulatory proteins. It entails tagging proteins with ubiquitin, a tiny regulatory protein that targets them for breakdown by the 26S proteasome, a large proteolytic complex. The ubiquitination process is extremely selective and comprises a cascade of enzymatic activities, including the activation of ubiquitin by the E1 enzyme, its conjugation with the target protein by the E2 enzyme, and the transfer of ubiquitin to the substrate by the E3 ubiquitin ligase (Pollard et al., 2024). This process can be repeated multiple times, resulting in the development of a polyubiquitin chain that the proteasome recognizes. When the substrate is recognized, it is unfolded and transported to the core of the proteasome, where it is degraded into short peptides, which are then further broken down into amino acids for cellular reuse (Deretic et al., 2013).

The ubiquitin-proteasome system plays an important role in the regulation of several cellular processes, including signal transmission, cell cycle control, and gene expression. For example, cyclin breakdown by the proteasome is required for the cell cycle to progress. Furthermore, the UPS removes misfolded proteins, preventing their accumulation, which could otherwise result in cellular malfunction and illnesses like dementia. Aberrations in the ubiquitin-proteasome system have been linked to a variety of illnesses, including cancer, where deregulation of this process can result in uncontrolled cell proliferation due to the stabilization of oncogenic proteins (Gomperts et al., 2009).

In contrast, the autophagy-lysosome pathway is a significant degradative mechanism for long-lived proteins, protein aggregates, and whole organelles. Autophagy is a tightly controlled process that involves the production of double-membraned vesicles called autophagosomes, which ingest cytoplasmic material and transport it to lysosomes for destruction (Kim et al., 2019). There are three forms of autophagy: macroautophagy, microautophagy, and chaperone-mediated autophagy (CMA). The most frequently researched of these is macroautophagy, which is commonly referred to as autophagy. This process begins with the formation of an isolation membrane, which expands to surround sections of the cytoplasm, resulting in the autophagosome. The autophagosome then fuses with the lysosome, where the contained material is broken down by lysosomal hydrolases.

Autophagy is essential for cellular homeostasis because it removes damaged organelles and protein aggregates, reducing cellular harm and preserving the quality of cellular components. It also plays a role in the cellular response to stressors such as food restriction, where it supplies an alternate source of energy by decomposing cytoplasmic components and recycling the resultant biomolecules. Furthermore, autophagy plays a role in immunological responses by eliminating intracellular pathogens and presenting antigens for immune identification (Deretic et al., 2013).

However, dysregulation of autophagy has been linked to a variety of illnesses, including neurological disorders, cancer, and cardiovascular disease. In neurodegenerative illnesses including Parkinson's and Alzheimer's, defective autophagy causes the formation of toxic protein aggregates, which contributes to neuronal death and disease progression (Nixon 2013). Autophagy, on the other hand, can have a dual role in cancer, operating as a tumor suppressor by reducing cell proliferation while also serving as a survival mechanism for cancer cells under metabolic stress, boosting tumor growth (Pollard et al., 2024). The context-dependent involvement of autophagy in cancer underlines the pathway's intricacy and therapeutic promise. The regulation of protein levels in cells is aided by other protein breakdown pathways apart from autophagy and the ubiquitin-proteasome system. A mechanism that selects misfolded proteins in the endoplasmic reticulum (ER) for proteasome degradation is the endoplasmic reticulum-associated degradation (ERAD) pathway. To avoid misfolded protein buildup, which can cause ER stress and activate the unfolded protein response (UPR), ERAD is essential for preserving ER homeostasis (Baynes and Dominiczak, 2009). Cell fate decisions are linked to ERAD because the UPR is a cellular stress response that might trigger apoptosis if the stress is not alleviated. Its primary goal is to restore ER function.

Apoptosis and caspase involvement in protein breakdown constitute another significant route. A family of cysteine proteases known as caspases is involved in the breakdown of cellular structures and ultimate cell death during apoptosis by cleaving particular substrates (Deretic et al., 2013). Caspases are largely recognized for their involvement in apoptosis, but they also play a function in non-apoptotic processes, such as the control of differentiation and inflammation, highlighting their significance in cell biology.

Protein breakdown mechanisms are closely regulated, with a complex network of signaling channels responding to numerous physiological stimuli. For example, the mammalian target of rapamycin (mTOR) pathway is a major regulator of autophagy, with activation inhibiting autophagy in response to nutrition availability and inhibition promoting autophagy in nutrient-deprived situations (Kim et al., 2011). Similarly, the nuclear factor erythroid 2-related factor 2 (Nrf2) pathway controls gene expression in the ubiquitin-proteasome system and autophagy, especially in response to oxidative stress (Zhang et al., 2015). The interaction of various signaling channels guarantees that protein breakdown is perfectly synchronized with cellular requirements and environmental conditions.

7.6. CHAPERONE PROTEINS

Chaperone proteins are crucial for helping to maintain cellular proteostasis, which is the process by which cells control protein production, folding, and destruction. These molecular chaperones aid in the correct folding of nascent polypeptides, prevent the aggregation of misfolded proteins, and help refold or destroy damaged proteins, ensuring that the proteome stays functioning and cellular homeostasis is preserved. Chaperone proteins are important because they participate in a variety of cellular activities, such as protein trafficking, signal transduction, and the stress response, as well as their role in many diseases when their function is impaired (Figure 7.5).

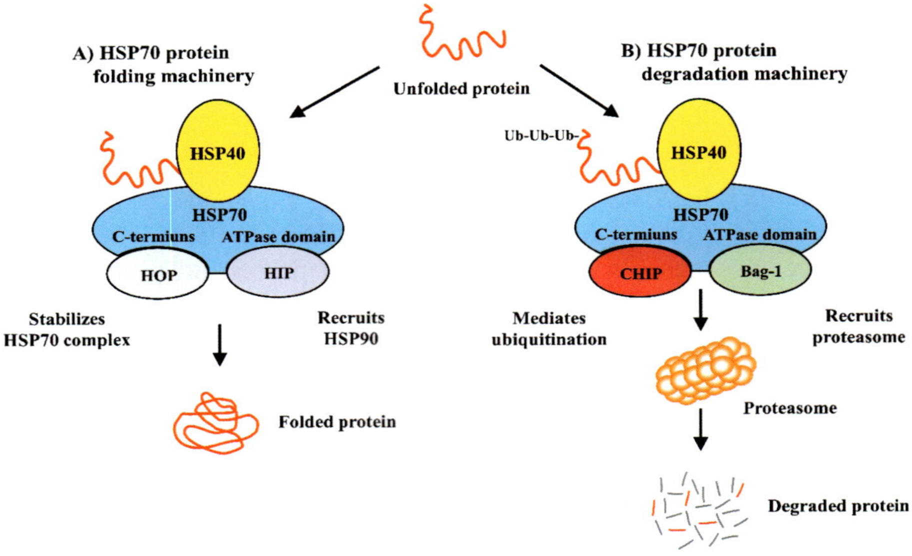

Figure 7.5. *The roles of chaperone proteins in cellular proteostasis.*

Source: Kim JY, Barua S, Huang MY, Park J, Yenari MA, Lee JE. (2020). Heat shock protein 70 (HSP70) induction: Chaperonotherapy for neuroprotection after brain injury. Cells, 9(9):2020. https://doi.org/10.3390/cells9092020).

The fundamental function of chaperone proteins is to help newly produced proteins fold properly. During protein synthesis, nascent polypeptide chains emerge from the ribosome in an unfolded state and must fold into the necessary three-dimensional conformations to function. Chaperones bind to these nascent chains, protecting hydrophobic areas from aggregation and regulating the folding process. Heat shock proteins (HSPs), particularly Hsp70, engage with nascent polypeptides and promote correct folding through ATP-driven cycles of binding and release. Hsp70 works with co-chaperones like Hsp40 and nucleotide exchange factors to increase its activity, demonstrating the complexity and coordination required in chaperone-assisted folding (Söti and Csermely, 2002).

Chaperones play an important role in the prevention of protein aggregation in addition to assisting in de novo folding. Misfolded proteins expose hydrophobic surfaces, resulting in the creation of toxic aggregates that have been linked to many diseases, including neurodegenerative disorders, such as Alzheimer's, Parkinson's, and Huntington's. Hsp90 and small heat shock proteins (sHSPs) bind to misfolded proteins, preventing them from aggregating and aiding their refolding or destruction via the ubiquitin-proteasome system or autophagy pathways. This action is critical for cellular survival under stress conditions, such as high temperatures, oxidative stress, or mutations in protein-coding genes, which increase the likelihood of protein misfolding and aggregation (Hartl et al., 2011).

Furthermore, chaperones help to degrade proteins that cannot be refolded, preventing damaged or irreversibly misfolded proteins from accumulating and causing cellular toxicity. This activity is especially crucial for controlling protein quality in organelles like the endoplasmic reticulum (ER), where freshly generated membrane and secretory proteins are folded. Chaperones are used in the ER-associated degradation (ERAD) pathway to identify misfolded proteins and target them for ubiquitination and proteasome destruction (Baynes and Dominiczak, 2009). Chaperones such as Binding Immunoglobulin Protein (BiP), also known as GRP78, play an important role in the ERAD process by binding to misfolded proteins and preventing them from aggregating within the ER lumen.

Chaperone proteins are also essential for the cellular stress response, especially during heat shock or other stressors that affect protein homeostasis. Under these conditions, cells increase the expression of heat shock proteins, which help to stabilize and refold broken proteins or target them for destruction (Hartl et al., 2011). Heat shock factors (HSFs) regulate this reaction by translocating to the nucleus when misfolded proteins are detected and activating heat shock gene transcription. The coordinated action of chaperones throughout the stress response is critical for cell survival, allowing cells to recover from proteotoxic stress and restore proteostasis.

Additionally, chaperones have a role in protein trafficking control and multiprotein complex formation. For example, Hsp70 and Hsp90 are known to help with the maturation and stabilization of steroid hormone receptors and other signaling proteins, guaranteeing their appropriate placement and function inside the cell (Pollard et al., 2024). Chaperones also help to assemble protein complexes by stabilizing individual subunits and avoiding premature or erroneous interactions. This activity is especially critical in the formation of large protein complexes like the proteasome and the mitochondrial electron transport chain, which require perfect assembly of numerous subunits to operate properly. Chaperone protein dysfunction has been linked to a variety of illnesses, underscoring their importance in maintaining cellular health. The inability of chaperone systems to regulate misfolded proteins in neurodegenerative illnesses results in the formation of toxic aggregates, which contribute to neuronal death and disease development (Soti and Csermely, 2002). Furthermore, several malignancies use chaperone mechanisms to help tumor cells grow and survive rapidly. Hsp90, for example, is frequently overexpressed in cancer cells, where it stabilizes a broad range of oncogenic proteins, making it a target for anti-cancer therapy. Mutations in chaperone proteins or their client proteins can cause metabolic illnesses like cystic fibrosis or Gaucher's disease by causing the misfolding and loss of activity of enzymes engaged in metabolic pathways (Soti and Csermely, 2002).

7.7. PROTEIN MISFOLDING DISEASES

Protein misfolding illnesses, also known as conformational diseases, are a class of conditions distinguished by the accumulation of improperly folded proteins within cells and tissues. These illnesses demonstrate the importance of protein folding in sustaining cellular function and the repercussions of folding defects in cell biochemistry. Proteins' biological activity relies on their appropriate folding into functional three-dimensional structures. When proteins fail to fold correctly, they can collect forming insoluble fibrils or plaques, interfering with normal cellular processes and contributing to disease progression.

Alzheimer's disease, a neurodegenerative condition caused by the accumulation of amyloid-beta (Aβ) peptides in the brain, is a well-known example of protein misfolding. These peptides are produced from the amyloid precursor protein (APP), which, when broken improperly, results in the production of Aβ. Alzheimer's disease is characterized by the accumulation of extracellular amyloid plaques caused by misfolded Aβ peptides. These plaques disrupt synaptic function, activate inflammatory responses, and contribute to neuronal death, resulting in cognitive decline and memory loss (Nusse and Clevers, 2017). Furthermore, hyperphosphorylation of the tau protein, another major hallmark of Alzheimer's, causes the production of neurofibrillary tangles, exacerbating the disease process.

Similarly, the misfolding and aggregation of the alpha-synuclein protein is linked to Parkinson's disease, another neurological condition. Alpha-synuclein is an intrinsically disordered protein that is involved in neurotransmitter release and synaptic vesicle trafficking in its native state. Lewy bodies, intracellular inclusions discovered in Parkinson's patients' brains, are created when alpha-synuclein misfolds and clumps into insoluble fibrils under pathological circumstances (Barbu et al., 2020). The distinctive motor symptoms of Parkinson's disease are caused by Lewy bodies, which are associated with neuronal malfunction and cell death, especially in dopaminergic neurons of the substantia nigra. Beyond neurodegenerative disorders, protein misfolding is linked to systemic amyloidosis, a category of diseases marked by the extracellular deposition of amyloid fibrils in numerous organs. These fibrils are made up of misfolded proteins such as immunoglobulin light chains, transthyretin, or serum amyloid A, depending on the type of amyloidosis. Amyloid fibril deposition affects normal tissue architecture and function, ultimately leading to organ failure. For example, in light-chain amyloidosis, the buildup of amyloid fibrils in the heart can cause restrictive cardiomyopathy, which is characterized by decreased cardiac function due to hardening of the heart walls (Baynes and Dominiczak, 2009). The biological mechanisms that ordinarily prevent protein misfolding include molecular chaperones and the ubiquitin-proteasome pathway. Molecular chaperones help nascent and stress-denatured proteins fold properly by avoiding aggregation and facilitating correct conformation. Chaperones, such as heat shock proteins (HSPs), are activated in response to physiological stress and play an important role in protein homeostasis, or proteostasis. Misfolded proteins that escape chaperone supervision are normally degraded by the ubiquitin-proteasome system. However, in many protein misfolding illnesses, these quality control systems are overloaded or compromised, resulting in the buildup of misfolded proteins and the development of hazardous aggregates (Pollard et al., 2024).

The harmful effects of protein aggregation are complex. Misfolded proteins can affect cellular function by trapping important physiological components, decreasing proteasome action, and causing endoplasmic reticulum (ER) stress. Misfolded proteins accumulate in the ER, triggering the unfolded protein response (UPR), a cellular stress response designed to restore ER function. However, chronic or severe ER stress can cause apoptosis, which contributes to cell death in protein-misfolding illnesses (Pollard et al., 2024).

Therapeutic approaches to protein misfolding illnesses have centered on improving the cellular capacity to manage misfolded proteins, either by increasing the activity of molecular chaperones, improving proteasome function, or encouraging autophagy, another cellular breakdown mechanism. Small compounds known as chemical chaperones can stabilize misfolded proteins, preventing them from aggregating and promoting proper folding (Hanahan and Weinberg, 2011). Furthermore, proteostasis regulators, which control the production of chaperones and proteasome components, are being investigated as potential treatments for restoring protein homeostasis in damaged cells.

Protein misfolding diseases constitute a serious issue because the buildup of misfolded proteins causes cellular malfunction and illness. Understanding the molecular mechanisms that control protein folding and cellular responses to misfolded proteins is critical for creating effective treatments for these severe diseases. As research in this field progresses, new therapeutic techniques targeting protein folding and degradation pathways show promise in alleviating the impact of protein misfolding illnesses.

Chapter 8

Cell Cycle and Cell Division

Contents

The cell cycle is a fundamental biological process that regulates cell growth, replication, and division. This process has four primary phases: G1, S, G2, and M. During the G1 phase, cells grow and prepare for DNA replication, which occurs in the S phase. In the G2 phase, cells continue to grow and prepare for mitosis, ensuring that all cellular components are adequately duplicated (Chagin et al., 2010). The M phase, or mitosis, is defined as the precise segregation of duplicated chromosomes into two daughter cells, followed by cytokinesis, which divides the cytoplasm and completes cell division (Alberts et al., 2015).

The regulation of the cell cycle is essential for maintaining cellular homeostasis and preventing uncontrolled proliferation, which can lead to oncogenesis. Cyclins and cyclin-dependent kinases (CDKs) regulate the cell cycle by forming complexes that control the progression through the various phases. Furthermore, checkpoints in the cycle ensure that cells do not move on to the next phase without completing critical tasks like DNA replication and damage repair (Pollard et al., 2022). Understanding the molecular mechanisms that govern the cell cycle is critical for progressing knowledge in cellular biology and developing therapeutic strategies for cancer and other proliferative disorders.

8.1. CELL CYCLE PHASES

The cell cycle is a highly regulated series of events that allows cells to grow, replicate their genetic material, and divide into daughters. This cycle is essential for cellular function, development, and proliferation in all living organisms. It includes several distinct phases: G1 (Gap 1), S (Synthesis), G2 (Gap 2), and M (Mitosis), each of which is distinguished by specific biochemical and structural activities that ensure proper cell division and genomic integrity (Branzei and Foiani, 2008) (Figure 8.1).

In the G1 phase, cells are primarily concerned with growth and metabolic activity. This phase is critical for the synthesis of RNA and proteins required for DNA replication and the subsequent stages of the cell cycle. During G1, the cell evaluates its surroundings, including the availability of nutrients and growth factors that are required for progression through the cycle. Cells that do not meet the criteria for further progression may enter a quiescent state known as G0, in which they are metabolically active but do not divide (Alberts et al., 2015). The restriction point, a checkpoint that determines whether conditions are favorable for DNA replication, is an important feature of G1. If this checkpoint is passed, the cell has committed to completing the cycle.

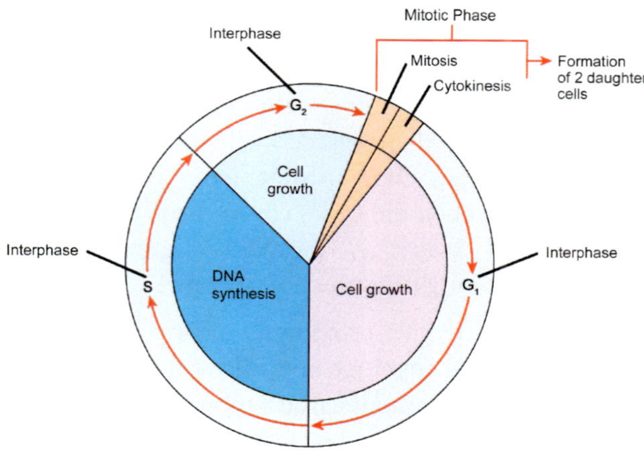

Figure 8.1. *Schematic view of different phases of the cell cycle.*

Source: https://courses.lumenlearning.com/suny-biology1/chapter/the-cell-cycle/.

Following the G1 phase, cells enter the S phase, which is when DNA synthesis occurs. This phase is distinguished by the replication of the cell's genetic material, which ensures that each daughter cell receives an identical copy of the DNA. DNA replication is a highly regulated and precise process in which helicases unwind the DNA double helix, replication forks are assembled, and new DNA strands are synthesized by DNA polymerases. During this phase, the entire genome is duplicated, with multiple replication origins activated to ensure completion on time. Furthermore, the synthesis of histone proteins, which package DNA into chromatin, is tightly coordinated with DNA replication to ensure proper chromosome structure and function (Alberts et al., 2015; Pollard et al., 2022).

The G2 phase occurs after DNA synthesis and acts as a second gap phase, allowing the cell to grow and prepare for mitosis. During G2, the cell continues to produce proteins and organelles, ensuring that it has the components required for division. This phase also includes a critical checkpoint that checks the integrity of the replicated DNA. Any damage or errors discovered during replication must be repaired before the cell can enter mitosis, ensuring the integrity of genetic information passed down to daughter cells. Cyclin-dependent kinases (CDKs) and their regulatory subunits, cyclins, tightly regulate the G2/M checkpoint, which controls the transition into mitosis by phosphorylating target proteins that initiate the mitotic program (Baynes and Dominiczak, 2009).

Mitosis, also known as the M phase, is the most active stage of the cell cycle, during which the cell's nucleus and cytoplasm divide to produce two genetically identical daughter cells. Mitosis is further divided into four stages: prophase, metaphase, anaphase, and telophase, each with distinct events that ensure accurate chromosome segregation. During prophase, chromatin condenses into visible chromosomes, and the mitotic spindle, made up of microtubules, starts to form. The nuclear envelope also degrades, allowing spindle fibers to connect to chromosomes at their kinetochores. Metaphase follows, during which the metaphase plate is formed when the cell's chromosomes align at the equatorial plane (Satyanarayana and Kaldis, 2009). This alignment ensures that each daughter cell receives one copy of each chromosome.

Anaphase is initiated by the proteolytic cleavage of cohesin proteins, which hold sister chromatids together and allow them to be pulled apart to opposite poles of the cell by spindle microtubules. This separation ensures that each daughter cell receives the same set of chromosomes. Telophase is the end of mitosis, when the separated chromatids, now considered individual chromosomes, reach the poles and begin to decondense. The nuclear envelope re-forms around each set of chromosomes, resulting in the restoration of two distinct nuclei within the cell (Chagin et al., 2010; Koutroumpas and Lygeros, 2011).

Cytokinesis is the final step in the cell cycle, dividing the parent cell's cytoplasm into two daughter cells. This process begins during telophase and ends shortly thereafter, ensuring that each daughter cell has a complete set of organelles and adequate cellular material to function independently. In animal cells, cytokinesis is the formation of a contractile ring made up of actin and myosin, which constricts the cell membrane creating two separate daughter cells (Satyanarayana and Kaldis, 2009; Alberts et al., 2015). Cytokinesis in plant cells occurs when a cell plate forms and develops into a new cell wall that separates the two daughter cells.

8.2. REGULATION OF CELL CYCLE PROGRESSION

The regulation of the cell cycle is a complex and tightly controlled process required for cellular integrity, growth, and division in all eukaryotic organisms. It is governed by a complex network of signaling pathways and regulatory proteins that ensure the cycle's orderly progression through its four phases: G1, S, G2, and M. This regulation is critical for preventing genomic instability and uncontrolled cell proliferation, which can lead to pathological conditions like cancer (Satyanarayana and Kaldis, 2009) (Figure 8.2).

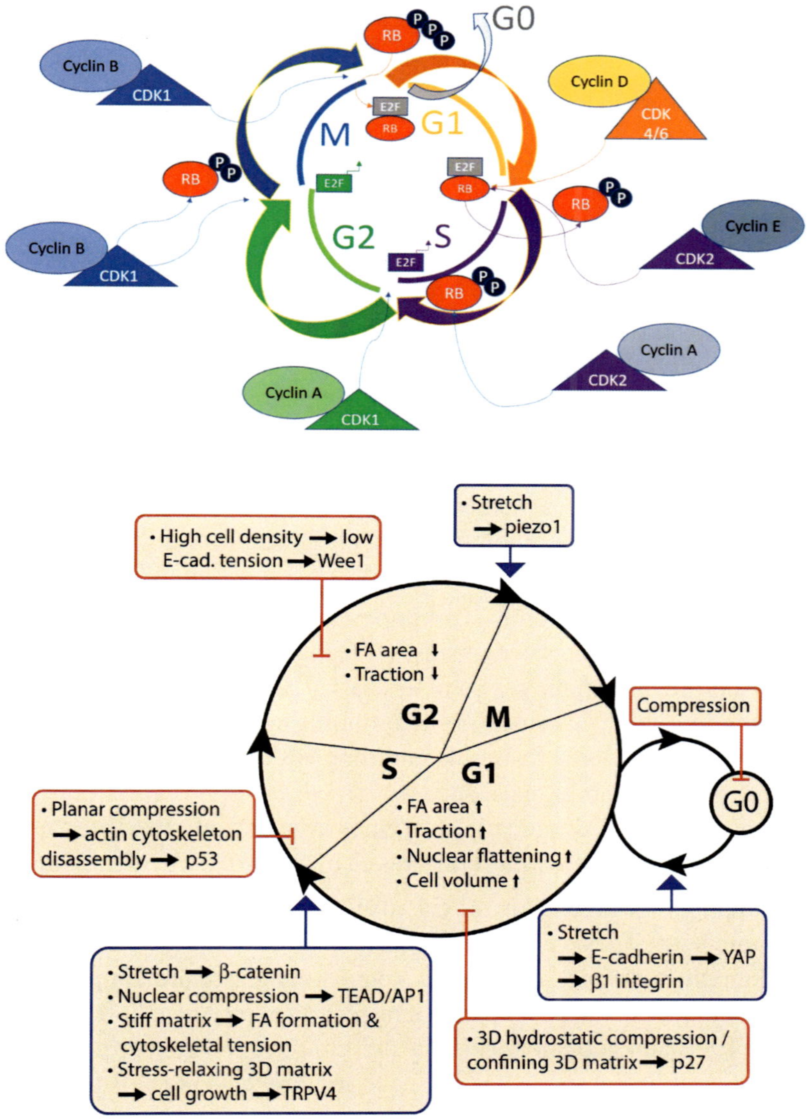

Figure 8.2. *The regulation of cell-cycle progression and division.*

Source: Javed A, Yarmohammadi M, Korkmaz KS, Rubio-Tomás T. (2023). The regulation of cyclins and cyclin-dependent kinases in the development of gastric cancer. International Journal of Molecular Sciences, 24(3):2848. https://doi.org/10.3390/ijms24032848).

Cyclins and cyclin-dependent kinases (CDKs) play an important role in cell cycle regulation. CDKs are serine/threonine kinases that, when activated by binding to specific cyclins, phosphorylate target proteins to propel the cell cycle forward. Cyclins are regulatory proteins whose concentrations vary throughout the cell cycle, influencing CDK activity at different stages. For example, during the G1 phase, cyclin D binds to CDK4/6, promoting the transition from G1 to S. Cyclin E associates with CDK2 to facilitate the G1/S transition, whereas cyclin A binds to CDK2 in S phase and CDK1 in G2, preparing the cell for mitosis. Finally, cyclin B binds to CDK1, which drives the cell into mitosis (Inzé and De Veylder, 2006; Alberts et al., 2015).

CDK inhibitors (CKIs), which act as critical cell cycle brakes, further regulate the activity of cyclin-CDK complexes. CKIs are classified into two families: the INK4 family, which inhibits only CDK4/6, and the Cip/Kip family, which inhibits a wider range of cyclin-CDK complexes. The progression of the cell through the cell cycle is determined by the balance of cyclin and CKI levels. For example, p21, a member of the Cip/Kip family, is upregulated in response to DNA damage, binding to and inhibiting cyclin-CDK complexes, thereby halting the cell cycle to allow for DNA repair (Pollard et al., 2022).

The presence of checkpoints is an important aspect of cell cycle regulation as they act as surveillance mechanisms, ensuring that each phase is completed adequately before the cell moves on to the next. The G1 checkpoint, also known as the restriction point, is critical in determining whether the cell has enough size, energy reserves, and a favorable environment for DNA replication. It also responds to extracellular signals, such as growth factors. If conditions are not favorable, cells can exit the cycle and enter a quiescent state known as G0, where they remain until conditions improve (Inzé and De Veylder, 2006).

DNA damage checkpoints in the G1 and G2 phases are critical for preserving genomic integrity. These checkpoints detect DNA damage and, if present, trigger a signaling cascade involving the tumor suppressor protein p53, which is an important player in the DNA damage response. In response to DNA damage, p53 is stabilized and activated, causing the transcription of target genes, such as p21, which inhibits cyclin-CDK activity and prevents the cell from moving on to the next phase. This pause allows the repair of damaged DNA before replication or mitosis, thus preventing the propagation of mutations (Boal, 2012; Pollard et al., 2022).

The S phase checkpoint monitors the integrity of DNA replication. If replication forks are stalled or DNA lesions are detected, the ATR kinase is activated, phosphorylating and activating downstream effectors such as Chk1. Activated Chk1 then inhibits CDC25 phosphatases, which prevents the activation of the cyclin A-CDK2 and cyclin B-CDK1 complexes, effectively stopping the cell cycle (Inzé and De Veylder, 2006). This mechanism ensures that DNA replication is complete and any errors are corrected prior to mitosis.

The G2/M checkpoint ensures that cells do not progress to mitosis with damaged or unreplicated DNA. The G2/M checkpoint, like the G1 checkpoint, is regulated by p53 and other signaling pathways involving kinases like ATM and ATR. These kinases activate Chk1 and Chk2, which inhibit CDC25C, a phosphatase required for the activation of the cyclin B-CDK1 complex. Inhibiting CDC25C prevents the dephosphorylation and activation of CDK1, thereby blocking the entry into mitosis until DNA damage is repaired (Pollard et al., 2022).

During mitosis, the spindle assembly checkpoint ensures that all chromosomes are properly attached to the spindle apparatus before anaphase begins. This checkpoint prevents chromosome missegregation and aneuploidy by blocking the anaphase-promoting complex/cyclosome (APC/C), a ubiquitin ligase that degrades securin and cyclin B. When chromosomes are not properly attached, the checkpoint proteins Mad2 and BubR1 activate and inhibit APC/C, preventing the onset of anaphase and allowing time for proper spindle attachment (Inzé and De Veylder, 2006; Satyanarayana and Kaldis, 2009).

Extracellular signals such as growth factors, cytokines, and cell adhesion molecules all have an effect on cell cycle regulation. Growth factors activate signaling pathways such as Ras/Raf/MEK/ERK and PI3K/AKT, which promote cell cycle progression by upregulating cyclins and downregulating CKIs. Furthermore, cell adhesion to the extracellular matrix (ECM) sends signals through integrins that can modulate cell cycle regulators, linking the cellular environment to cell cycle control (Baynes and Dominiczak, 2009).

Cancer is defined by the dysregulation of cell cycle control. Mutations in genes that encode cyclins, CDKs, CKIs, and other cell cycle regulators can result in uncontrolled cell proliferation. Overexpression of cyclin D1 or loss of p16^INK4a, a CKI, is common in cancers and can lead to uncontrolled progression through the G1 phase. Furthermore, mutations in the TP53 gene, which encodes the p53 protein, are commonly found in cancer and result in the loss of DNA damage checkpoints, allowing cells with genomic instability to proliferate (Satyanarayana and Kaldis, 2009).

Understanding the molecular mechanisms that underpin cell cycle regulation is critical for developing targeted cancer treatments. CDK inhibitors, such as palbociclib and ribociclib, have been developed and approved to treat cancers such as breast cancer, demonstrating the therapeutic potential of targeting cell cycle regulators (Alberts et al., 2015). These drugs inhibit CDK4/6, which prevents phosphorylation of the retinoblastoma protein (Rb), halting cell cycle progression at the G1 checkpoint in cancer cells.

8.3. CHECKPOINTS IN THE CELL CYCLE

Cell cycle checkpoints are important control mechanisms that ensure the accuracy and integrity of cellular division. These checkpoints act as surveillance systems, monitoring and regulating the progression of the cell cycle, ensuring that each phase is properly completed before the next begins. They play a critical role in maintaining genomic stability, preventing the spread of damaged DNA, and preserving cellular homeostasis. The major checkpoints in the cell cycle are located at the G1/S transition, S phase, G2/M transition, and metaphase-anaphase transition during mitosis, respectively. Each checkpoint involves a complex interplay of signaling pathways and regulatory proteins, orchestrating a highly controlled sequence of events essential for cellular function and viability (Satyanarayana and Kaldis, 2009).

The G1/S checkpoint, also known as the restriction point in mammalian cells, is a critical control mechanism that determines whether a cell commits to DNA replication and subsequent division or exits the cell cycle into the quiescent state known as G0. This checkpoint assesses the DNA's integrity and the cellular environment, including the presence of sufficient nutrients, growth factors, and energy reserves. The retinoblastoma protein (Rb) is a key regulatory player in this checkpoint because it binds to E2F transcription factors when hypophosphorylated, preventing the transcription of genes required for S phase entry. The activation of cyclin D-CDK4/6 complexes causes phosphorylation of Rb, resulting in the release of E2F and transcription of S phase-promoting genes. In response to DNA damage or unfavorable conditions, the tumor suppressor protein p53 can induce the expression of the cyclin-dependent kinase inhibitor (CKI) p21, which inhibits CDK activity and halts cell cycle progression (Alberts et al., 2015; Wenzel and Singh, 2018).

The S phase checkpoint's primary function is to monitor DNA replication. During this phase, the entire genome of the cell is duplicated, which must be done with high accuracy to avoid mutations and chromosomal abnormalities. The S phase checkpoint ensures that DNA replication is both complete and accurate. It tracks replication fork progression and responds to any DNA damage that occurs during replication. This checkpoint is heavily influenced by the ataxia-telangiectasia and Rad3-related protein (ATR). ATR is activated in response to single-stranded DNA regions formed at stalled replication forks, phosphorylating, and activating checkpoint kinase 1 (Chk1). Chk1 then inhibits CDC25 phosphatases, preventing the activation of cyclin A-CDK2 complexes, thereby stalling the cell cycle and providing time for DNA repair and completion of replication. This mechanism is vital for preventing replication stress and ensuring genomic stability (Pollard et al., 2022). The G2/M checkpoint is a critical barrier between the end of DNA replication and the start of mitosis. This checkpoint ensures that all DNA has been correctly replicated and that any DNA damage caused by replication has been repaired before the cell enters mitosis. Failure to complete these steps can lead to genomic instability, aneuploidy, and tumorigenesis. The G2/M checkpoint, like the G1/S checkpoint, is regulated by a complex network of proteins, including the ATM and ATR kinases, which are activated in response to DNA damage. These kinases phosphorylate a variety of substrates, including Chk1 and Chk2, which then phosphorylate and inhibit the CDC25C phosphatase. By inhibiting CDC25C, these checkpoints prevent the activation of cyclin B-CDK1 complexes, halting the cell cycle in G2 and allowing time for DNA repair. Furthermore, the tumor suppressor protein p53 plays a critical role in the G2/M checkpoint by inducing the transcription of p21, which inhibits cyclin-CDK complexes, thereby preventing cell cycle progression (Boal, 2012; Barnum and O'Connell, 2014).

The mitotic checkpoint, also known as the spindle assembly checkpoint (SAC), is a mechanism that regulates chromosome segregation during mitosis. This checkpoint is active during the metaphase-anaphase transition, monitoring the attachment of spindle microtubules to kinetochores, which are protein complexes found at the centromeres of chromosomes. The SAC ensures that all chromosomes are correctly attached to the spindle apparatus and under proper tension before anaphase begins. This checkpoint prevents aneuploidy, which occurs when cells have an abnormal number of chromosomes and can result in cell death or cancerous transformation. The SAC is based on a set of conserved proteins, including Mad2, BubR1, and Mps1, which are recruited to unattached kinetochores. These proteins form the mitotic checkpoint complex (MCC), which inhibits the anaphase-promoting complex/cyclosome (APC/C), a ubiquitin ligase that degrades securin and cyclin B. APC/C inhibition prevents sister chromatid separation and anaphase progression until all chromosomes are correctly attached to the spindle microtubules (Barnum and O'Connell, 2014; Pollard et al., 2022).

In addition to these well-defined checkpoints, several auxiliary pathways and feedback mechanisms help to fine-tune cell cycle control. For example, DNA damage response (DDR) pathways, which involve a variety of proteins such as BRCA1, Rad51, and FANCD2, work with cell cycle checkpoints to repair DNA and maintain genome integrity. These pathways can activate or amplify checkpoint signaling, resulting in additional layers of regulation. Additionally, the interaction between cyclin-dependent kinases (CDKs) and their inhibitors (CKIs) such as p21, p27, and p16 is critical in fine-tuning cell cycle transitions. By inhibiting CDKs, CKIs ensure that cells do not advance prematurely or inappropriately through the cycle, which is especially important in response to DNA damage or other cellular stresses (Wenzel and Singh, 2018).

External signals such as growth factors, cytokines, and extracellular matrix interactions can also modulate cell cycle progression via various signal transduction pathways, influencing cell cycle checkpoint regulation. For example, the Ras/MAPK and PI3K/AKT pathways are known to promote cell cycle progression by increasing cyclin D expression and CDK4/6 activity, which aids in the G1/S transition. Antiproliferative signals like those mediated by transforming growth factor-beta (TGF-β) can inhibit cell cycle progression in G1 by inducing CKI expression (Barnum and O'Connell, 2014; Wenzel and Singh, 2018). The interaction between extracellular signals and cell cycle regulators connects the cellular environment to the intrinsic machinery that governs cell proliferation.

The dysregulation of cell cycle checkpoints is a hallmark of cancer. Mutations or alterations in checkpoint proteins such as p53, ATM, ATR, and SAC components can result in unchecked cell cycle progression, genomic instability, and malignant transformation. Mutations in TP53, the gene that encodes p53, are one of the most common alterations in human cancers, resulting in the loss of G1/S and G2/M checkpoint control as well as the accumulation of genetic mutations. Similarly, overexpression of cyclin D1 or loss of CKIs such as p16 can disrupt G1 checkpoint control, leading to uncontrolled cell proliferation (Satyanarayana and Kaldis, 2009; Pollard et al., 2022). Understanding the molecular mechanisms underlying checkpoint regulation is important for cancer therapy since many chemotherapeutic agents and targeted therapies target defects in checkpoint control to selectively kill cancer cells.

8.4. MITOSIS AND CYTOKINESIS

Mitosis and cytokinesis are the final stages of the cell cycle, where chromosomes are precisely segregated and the cytoplasm is divided to produce two genetically identical daughter cells. These processes are critical for cellular proliferation and tissue development in eukaryotic organisms, ensuring that each daughter cell receives an accurate and complete set of genetic information. Mitosis is a highly orchestrated series of events that can be divided into five phases: prophase, prometaphase, metaphase, anaphase, and telophase; cytokinesis is the physical division of the cell's cytoplasm (Pollard et al., 2022).

During prophase, the loosely organized chromatin fibers in the nucleus condense into visible chromosomes, each made up of two sister chromatids joined at the centromere. This condensation is facilitated by condensin complexes and is required for proper chromosome segregation during mitosis. Concurrently, the mitotic spindle, a dynamic structure made up of microtubules, starts to form outside the nucleus. The centrosomes, which serve as the microtubule-organizing centers, begin moving to opposite poles of the cell, establishing the bipolar spindle required for chromosome segregation (Hayashi and Karlseder, 2013).

Prometaphase occurs after prophase and is distinguished by the breakdown of the nuclear envelope, which allows spindle microtubules to access chromosomes. Mitotic kinases, such as cyclin-dependent kinase 1 (CDK1) phosphorylate nuclear Lamins and other nuclear envelope components facilitate this disassembly. Spindle microtubules can attach to kinetochores, protein complexes found on chromosome centromeres after the nuclear envelope is disassembled. This attachment is critical to chromosome alignment and segregation. Kinetochores serve a dual role by serving as attachment sites for microtubules and signaling platforms that generate the spindle assembly checkpoint (SAC) signals, ensuring that all chromosomes are properly attached to the spindle before proceeding to anaphase (Alberts et al., 2015).

Metaphase is the phase in which the chromosomes align at the cell's equatorial plane, also known as the metaphase plate. During this phase, the spindle assembly checkpoint becomes fully active, monitoring the attachment of spindle fibers to kinetochores and ensuring that each chromosome is bi-oriented, meaning its sister chromatids are attached to opposite poles. This checkpoint prevents premature separation of chromosomes, safeguarding against aneuploidy, which can cause cell death or disease states such as cancer (Hayashi and Karlseder, 2013).

Once all chromosomes are properly aligned and attached, the cell enters anaphase, which is distinguished by the separation of sister chromatids. Anaphase is triggered by the activation of the anaphase-promoting complex/cyclosome (APC/C), an E3 ubiquitin ligase that degrades securin, a protein that inhibits separase. After securin is degraded, separase is activated and cleaves the cohesin complexes that hold sister chromatids together, allowing them to be pulled apart toward opposite spindle poles by shortening kinetochore microtubules (Wenzel and Singh, 2018). This separation ensures that each daughter cell inherits an identical set of chromosomes.

Telophase is the final stage of mitosis, when the separated chromatids, which are now individual chromosomes, reach the cell's poles and begin to decondense back into chromatin. The nuclear envelope reassembles around each set of chromosomes, resulting in two distinct nuclei. This nuclear envelope reformation is necessary for reestablishing nuclear architecture and initiating gene expression in daughter cells. The mitotic spindle disassembles, and the cell gets ready for cytokinesis (Hayashi and Karlseder, 2013; Pollard et al., 2022).

Cytokinesis is the physical process by which the parent cell's cytoplasm divides into two daughter cells. In animal cells, cytokinesis begins with the formation of a contractile ring made up of actin filaments and myosin II that form beneath the plasma membrane at the metaphase plate. The contraction of this actomyosin ring, caused by the sliding of actin filaments past myosin, creates the force required to pinch the cell into two separate entities, a process known as cleavage furrow formation. This action is tightly regulated by small GTPases such as RhoA, which coordinates the assembly and contraction of the contractile ring (Baynes and Dominiczak, 2009).

Cytokinesis occurs in plant cells via a different mechanism due to the presence of a rigid cell wall. Plant cells undergo cytokinesis by forming a cell plate in the cell's center, rather than a contractile ring. The cell plate is made up of vesicles from the Golgi apparatus that combine at the point of division to form a new cell wall that separates the two daughter cells. The phragmoplast, a plant-specific structure composed of microtubules, actin filaments, and associated proteins, coordinates this process by guiding the vesicles to the cell plate assembly site (Boal, 2012; Hayashi and Karlseder, 2013).

8.5. MEIOSIS AND GAMETE FORMATION

Meiosis is an important mechanism that enables the creation of haploid gametes, which are required for sexual reproduction in eukaryotes. This highly regulated process consists of two rounds of cell division, known as meiosis I and meiosis II, after a single round of DNA replication. The diligent orchestration of these divisions reduces the number of chromosomes by half, ensuring genomic integrity over generations. The biochemical activities that drive meiosis are complex, including a sequence of coordinated interactions among enzymes, structural proteins, and regulatory factors.

During meiosis I, homologous chromosomes undergo pairing and recombination, both of which are essential for genetic variation. The synaptonemal complex, a proteinaceous structure that connects homologous chromosomes, facilitates this pairing, also known as synapsis. Recombination, which includes the exchange of genetic material across homologous chromosomes, is mediated by a complex of proteins, including the endonuclease Spo11, which causes double-strand breaks in the DNA (Van Spronsen et al., 2021). These breaks are then repaired via homologous recombination, a process that involves the alignment of homologous sequences and the development of crossovers, which exchange portions of DNA between chromosomes. This exchange enhances genetic variety and maintains normal chromosome segregation by creating physical linkages between homologs (Pollard et al., 2024).

The effective segregation of homologous chromosomes during meiosis I is an essential component of gamete development. The meiotic spindle, a dynamic structure made up of microtubules and related proteins, drives the process. The spindle apparatus connects to kinetochores, protein complexes formed at the centromeres of chromosomes, and aids their migration to opposing poles of the cell. Cohesins, protein complexes that hold sister chromatids together, are important in this process because they ensure that only homologous chromosomes are split during meiosis I while sister chromatids stay linked (Hunter, 2015). Following homolog separation, cytokinesis occurs, dividing the cell into two daughter cells, each with half the number of chromosomes.

Meiosis II, like mitosis, involves the separation and distribution of sister chromatids into distinct cells. Unlike mitosis, meiosis II does not include a previous cycle of DNA replication, resulting in the creation of four genetically different haploid gametes from a single diploid progenitor cell. This process is carefully regulated by a network of cell cycle proteins, including cyclins and cyclin-dependent kinases (CDKs), which ensures that meiosis proceeds in an orderly manner. Errors in meiosis regulation can result in aneuploidy, a disease marked by an aberrant number of chromosomes that is a leading cause of miscarriage and congenital abnormalities such as Down syndrome (Hunter, 2015).

The biology of gamete production is equally intricate, requiring extensive cellular reorganization. Males undergo spermatogenesis in the testes, which leads to the continual generation of sperm cells into maturity. Spermatogenesis consists of three primary stages: spermatogonial growth, meiosis, and spermiogenesis. During spermiogenesis, spermatids go through many morphological and biochemical changes, such as chromatin condensation, acrosome development, and flagellum creation, all of which are required for sperm motility and capacity to fertilize an egg (O'Donnell, 2014). These alterations are governed by a variety of physiological cues, including hormones like testosterone, which binds to androgen receptors and triggers a cascade of gene expression changes required for sperm production.

In females, oogenesis occurs in the ovaries and is defined by the creation of a limited number of oocytes, each of which is arrested in prophase I until recruited for ovulation. Unlike spermatogenesis, which is continuous, oogenesis is discontinuous, with meiosis only completing after fertilization. High levels of cyclic AMP (cAMP) maintain oocyte arrest in prophase I by inhibiting the activation of the maturation-promoting factor (MPF), a major regulator of meiotic development. cAMP levels decrease in response to hormonal stimulation during the menstrual cycle, allowing MPF to activate and drive the egg through the remaining phases of meiosis. The above process culminates in the formation of mature egg and polar bodies, which are byproducts of the asymmetric divisions that occur during oogenesis (Hunter, 2015).

The molecular pathways that control meiosis and gamete creation are remarkably conserved across species, highlighting their critical role in biology. For example, the role of cyclins and CDKs in cell cycle control, as well as the mechanism of homologous recombination in fostering genetic variety, are conserved. The evolutionary preservation of these routes demonstrates their crucial role in maintaining the effective transmission of genetic material across generations.

Research into the molecular mechanisms underpinning meiosis and gamete production has crucial implications for human fertility and infertility treatments. Advances in our understanding of these processes have resulted in the creation of assisted reproductive technologies (ART), such as in vitro fertilization (IVF), which rely on gamete manipulation for successful fertilization (Finn et al., 2016; O'Donnell, 2014). Furthermore, the study of meiotic mistakes has shed light on the causes of aneuploidy and informed tactics for screening and preventing chromosomal abnormalities in embryos.

Thus, meiosis and gamete creation are critical processes that assure the development of haploid gametes required for sexual reproduction. These processes are controlled by a complex network of molecular pathways that govern chromosome segregation, cell division, and cellular remodeling. Understanding the biochemistry of meiosis and gamete generation is critical for expanding our understanding of reproductive biology, with important implications for infertility therapy and genetic problem prevention.

8.6. CELL CYCLE CONTROL IN CANCER

Cell cycle regulation is critical in maintaining cellular homeostasis and preventing uncontrolled cell multiplication, which is characteristic of cancer. The cell cycle is made up of several closely regulated phases that ensure proper DNA replication and division. Disruptions in cell cycle regulatory pathways can result in genomic instability and oncogenesis, emphasizing the importance of understanding cell cycle control in the context of cancer biology and cell biochemistry.

CDKs and their regulatory partners, cyclins, play critical roles in controlling cell cycle progression. CDKs are serine/threonine kinases that phosphorylate target proteins after being activated by binding to particular cyclins, allowing the cell to progress through the cell cycle. For example, the cyclin D-CDK4/6 complex promotes DNA synthesis by phosphorylating the retinoblastoma protein (Rb) and releasing E2F transcription factors. Overexpression or amplification of cyclin D and CDK4/6 has been found in cancer, leading to uncontrolled cell proliferation caused by the absence of normal cell cycle checkpoints (Finn et al., 2016).

Tumor suppressor proteins such as p53 and Rb are essential for enforcing cell cycle checkpoints and inhibiting the spread of damaged DNA. p53, sometimes known as the "guardian of the genome," responds to DNA damage by initiating cell cycle arrest, allowing for DNA repair, or initiating apoptosis if the damage is irreversible. Mutations in the TP53 gene, which encodes p53, are among the most prevalent in human malignancies, and they cause the loss of these protective mechanisms, allowing tumors to form and advance. Similarly, mutations or deletions that inactivate the Rb pathway impair G1/S checkpoint control, contributing to uncontrolled cell division and oncogenesis (Jain et al., 2021).

Cell cycle inhibitors, such as the INK4 and CIP/KIP families, act as negative regulators by binding and inhibiting CDKs. INK4 family members, including p16^INK4a, block CDK4/6, which prevents cyclin D binding and Rb phosphorylation. CIP/KIP proteins, such as p21^Cip1 and p27^Kip1, have broad specificity and can inhibit several cyclin-CDK complexes at various cell cycle phases. In many malignancies, these inhibitors are downregulated or functionally inactivated, resulting in uncontrolled CDK activity and rapid cell cycle progression (Finn et al., 2016). The silencing of p16^INK4a by promoter hypermethylation is a common event in many malignancies, indicating its function in tumor suppression (Jain et al., 2021).

Deregulation of cell cycle checkpoints is exacerbated by abnormal signaling pathways that impact cyclin and CDK activity. Hyperactivation of the PI3K/Akt pathway through mutations or overexpression can accelerate cell cycle progression by raising cyclin D1 production and decreasing p27^Kip1 activity. Furthermore, the Ras/MAPK pathway promotes cyclin D1 transcription and aids in the progression through the G1 phase, and its constitutive activation is common in malignancies (Finn et al., 2016). Targeting these upstream signaling pathways has emerged as a potential therapeutic method for restoring normal cell cycle control in cancer cells.

DNA damage response (DDR) mechanisms are inextricably related to cell cycle regulation, protecting genomic integrity by synchronizing DNA repair processes with cell cycle arrest. When DNA damage is detected, key DDR kinases such as ATM and ATR activate checkpoint kinases Chk1 and Chk2, resulting in p53 stabilization and activation and cell cycle arrest. Deficiencies in DDR components impair the cell's ability to respond to genomic shocks, leading to mutation accumulation and an increased risk of cancer. ATM mutations, for example, have been linked to ataxia-telangiectasia, a disease with a susceptibility to lymphoid malignancies (Finn et al., 2016). Therapeutic interventions aimed at cell cycle regulators have shown potential in cancer treatment. CDK inhibitors like palbociclib and ribociclib have been designed to precisely block CDK4/6 activity, successfully arresting cancer cells in the G1 phase and showing efficacy in hormone receptor-positive breast tumors (Finn et al., 2015). Furthermore, using synthetic lethality to target DDR pathways, as demonstrated with PARP inhibitors in BRCA-mutated malignancies, demonstrates the therapeutic potential of modifying cell cycle regulatory systems (Pollard et al., 2024).

8.7. STEM CELLS AND REGENERATIVE MEDICINE

Stem cells and regenerative medicine constitute a new frontier in cell biochemistry, with far-reaching consequences for human health. These domains are inextricably linked with the principles of cell biochemistry, which give the underlying understanding required to alter cells for therapeutic purposes. Stem cells' unique ability to differentiate into several cell types underlines their potential in regenerative medicine, where they can be used to repair or replace damaged tissues and organs. This potential is based on the biochemical pathways that govern cell differentiation, proliferation, and maintenance, all of which are important in stem cell therapy and the larger area of regenerative medicine.

Stem cells are distinguished by their ability to self-renew and specialize into several cell lineages, which is inextricably related to the regulation of certain biochemical pathways. Pluripotent stem cells, such as embryonic stem cells (ESCs) and induced pluripotent stem cells (iPSCs), may differentiate into any cell type in the body, making them extremely beneficial in regenerative medicine. A network of transcription factors, including OCT4, SOX2, and NANOG, maintains pluripotency by regulating the expression of genes required for an undifferentiated state. The interaction of these transcription factors with epigenetic changes like DNA methylation and histone acetylation improves stem cells' ability to stay pluripotent or differentiate into specific cell types (Diamante and Martello, 2022).

Cell biochemistry is also important in the differentiation of stem cells, which is driven by a combination of intrinsic and extrinsic signals. Intrinsic factors include the aforementioned transcription factors and signaling molecules, whereas extrinsic factors include the cellular microenvironment, or niche, which provides biochemical cues that direct differentiation. For example, the Wnt signaling pathway, which is essential for embryonic development, has been demonstrated to influence stem cell fate by regulating the expression of genes involved in cell cycle regulation and differentiation (Jin et al., 2016).

The balance of Wnt signaling vs other pathways, such as Notch and Hedgehog, affects whether a stem cell remains undifferentiated or commits to a certain lineage.

Understanding and harnessing these metabolic pathways is critical in the development of effective regenerative medicine therapy. One interesting use is the use of stem cells to restore heart tissue after a myocardial infarction. The heart, as a mostly non-regenerative organ, has severe function loss following myocardial infarction as a result of cardiomyocyte death. Research has revealed that stem cells, specifically iPSCs, can be guided to differentiate into cardiomyocytes by manipulating biochemical pathways, such as activating the GATA4, MEF2C, and TBX5 transcription factors, all of which are essential for cardiac development. These cells can subsequently be implanted into the injured heart tissue, integrating with the host myocardium and restoring cardiac function (Matsui, 2016).

The generation of bioengineered tissues is another area where stem cells, regenerative medicine, and cell biochemistry meet. Tissue engineering uses cells, including stem cells, and biocompatible scaffolds to generate functional tissue structures. This approach's effectiveness is strongly dependent on knowing the molecular signals that control cell adhesion, proliferation, and differentiation on the scaffold. Growth factors like TGF-β and FGF, when combined with scaffold materials, can help differentiate stem cells into specific cell types like chondrocytes for cartilage repair or osteoblasts for bone regeneration (Pollard et al., 2024). The scaffold not only provides structural support, but it also mimics the extracellular matrix, supplying biochemical cues required for the created tissue's proper spatial organization and functionality.

Furthermore, the biochemistry of stem cells includes their metabolism, which varies significantly throughout differentiation. Stem cells primarily use glycolysis for energy production, a phenomenon known as the Warburg effect, which is also seen in cancer cells. However, as stem cells differentiate, they switch to oxidative phosphorylation, a more efficient mode of energy production (Zhang et al., 2016). This metabolic shift is closely controlled by many biochemical mechanisms, including the mTOR and AMPK pathways, which detect cellular energy levels and change metabolic activities accordingly. Understanding these metabolic alterations is critical for enhancing differentiation and increasing the efficacy of stem cell-based therapies.

In the context of neurodegenerative illnesses like Parkinson's and Alzheimer's, stem cells can replace lost or damaged neurons. The differentiation of stem cells into neurons is a complex process that necessitates precise control of biochemical pathways, such as those involving the neurogenin and

NeuroD transcription factors, which are essential for neurogenesis (Diamante and Martello, 2022). Furthermore, the significance of epigenetic control in neural differentiation is becoming more widely acknowledged, with changes like histone methylation and acetylation playing critical roles in activating or repressing neuronal development-related genes. By modifying these metabolic processes, researchers hope to create specific neuronal subtypes that can be transplanted into patients, potentially slowing or reversing neurodegenerative disease progression (Matsui, 2016).

Chapter 9

Membrane Transport and Cell Signaling

Contents

Membrane transport and cell signaling are important processes that help to maintain homeostasis and transmit signals within and between cells. Membrane transport is the movement of ions, molecules, and other substances across the cellular membrane, which is selectively permeable and primarily made up of a phospholipid bilayer. Transport mechanisms are classified as passive, relying on concentration gradients (e.g., diffusion and facilitated diffusion) or active, requiring energy input to move substances against their gradients (e.g., primary and secondary active transport) (Pollard et al., 2022).

Cell signaling, on the other hand, refers to the mechanisms by which cells perceive and respond to their external environment, coordinating complex cellular activities. This process frequently begins with the binding of signaling molecules, or ligands, to specific cell surface receptors, which triggers a series of intracellular events that transmit the signal to various cellular targets. These signaling pathways, such as the MAPK and PI3K-Akt pathways, are critical for controlling a variety of cellular functions, including growth, differentiation, and apoptosis (Alberts et al., 2015). Understanding these biochemical processes is critical for determining cellular function and developing therapeutic interventions for diseases involving dysregulated transport and signaling pathways.

9.1. PASSIVE AND ACTIVE TRANSPORT MECHANISMS

Cellular transport mechanisms are essential for maintaining cellular homeostasis and performing a variety of physiological functions. These transportation mechanisms are divided into two categories: passive and active transport. Both types are required to regulate the intracellular environment, maintain proper cell function, and facilitate communication between cells and their surroundings (Pollard et al., 2022).

Passive transport refers to the movement of molecules across a cell membrane without the use of cellular energy (ATP). This process is based on the kinetic energy of molecules and the concentration gradients that exist across the membrane. The three main types of passive transport are simple diffusion, facilitated diffusion, and osmosis. Simple diffusion occurs when small, nonpolar molecules like oxygen and carbon dioxide move directly across the lipid bilayer from higher to lower concentrations (Alberts et al., 2015). This movement continues until equilibrium is achieved when there is no net movement of molecules.

Another type of passive transport is facilitated diffusion, which involves the movement of larger or more polar molecules that cannot freely cross the lipid bilayers. This process necessitates the use of specific transmembrane proteins, such as carrier and channel proteins, to help molecules move down their concentration gradient. Channel proteins form pores that allow ions and small molecules to pass through, whereas carrier proteins change conformation to transport molecules across membranes. Facilitated diffusion can be seen in the transport of glucose into cells via GLUT transporters. Osmosis, a type of facilitated diffusion, refers to the passive movement of water molecules across a semipermeable membrane. Water moves from areas of low solute concentration to areas of high solute concentration, aiming to equalize solute concentrations on both sides of the membrane (Boal, 2012).

Active transport, on the other hand, requires the direct use of energy, usually in the form of ATP, to move molecules against their concentration gradient. This energy-dependent process enables cells to maintain concentration gradients, which are necessary for a variety of cellular functions. There are two types of active transport: primary and secondary.

Primary active transport involves direct hydrolysis of ATP to provide the energy required for the transport of molecules across the membrane. The sodium-potassium pump (Na+/K+ ATPase) is a well-known example of primary active transport. It is responsible for maintaining the electrochemical gradients of sodium and potassium ions across the plasma membrane. The Na+/K+ ATPase pumps three sodium ions out of the cell and two potassium ions into the cell, creating and maintaining the membrane potential required for nerve impulse transmission and muscle contraction (Forrest et al., 2011; Alberts et al., 2015).

Secondary active transport, also known as co-transport, does not use ATP directly to move molecules. Instead, it relies on the energy stored in the electrochemical gradient produced by primary active transport. Secondary active transport is further divided into symport and antiport mechanisms. Symporters transport two or more molecules in the same direction across a membrane, whereas antiporters transport molecules in opposite directions. The sodium-glucose cotransporter (SGLT) is an example of symport, as it uses the sodium gradient established by the Na+/K+ ATPase to drive glucose uptake into cells against its concentration gradient (Stein, 2012). On the other hand, the sodium-calcium exchanger (NCX) is an example of an antiporter, where the inward flow of sodium ions is coupled with the outward flow of calcium ions, helping regulate intracellular calcium levels.

Understanding the mechanisms of passive and active transport is important in cell biology because it explains how cells regulate their internal environment, communicate with their surroundings, and maintain homeostasis. Disruptions in these transport mechanisms can result in a variety of diseases and disorders. Mutations in genes encoding ion channels or transporters, for example, can cause cystic fibrosis, which causes thick mucus buildup in the lungs due to defective chloride ion transport. Similarly, abnormalities in glucose transport can contribute to the pathophysiology of diabetes mellitus, as impaired glucose uptake by cells leads to elevated blood glucose levels (Alberts et al., 2015; Forrest et al., 2011).

9.2. ION CHANNELS AND PUMPS

Membrane ion channels and pumps are essential components of cellular membranes, helping to regulate ion gradients and transport substances across them. These proteins are responsible for cellular homeostasis, electrical excitability, and signal transduction, all of which are required for a variety of physiological functions. The biochemistry of these membrane proteins involves intricate mechanisms, which have been extensively studied to understand their role in cellular function and their implications in various diseases (Stein, 2012) (Figure 9.1).

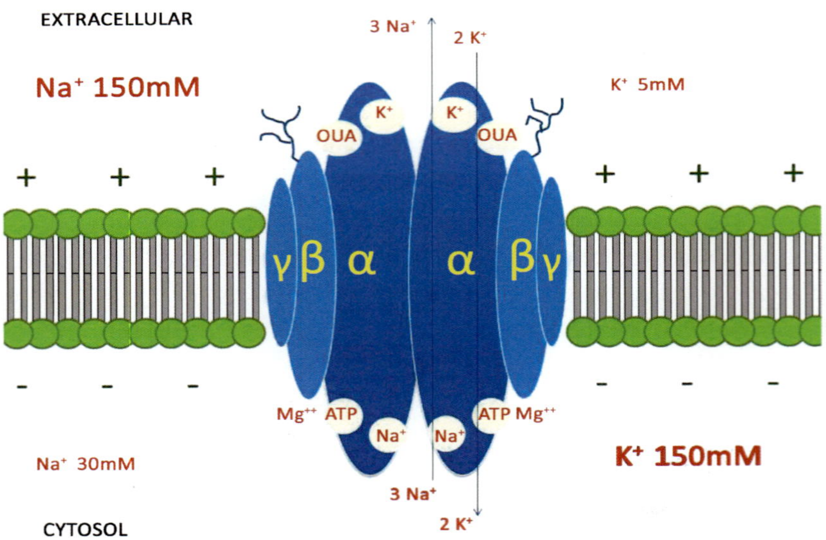

Figure 9.1. *Sketch of the structure of the cell membrane showing the ion gradient, ion channel, and ion pumps for exchanging sodium and potassium ions.*

Source: Felippe Gonçalves-de-Albuquerque C, Ribeiro Silva A, Ignácio da Silva C, Caire Castro-Faria-Neto H, Burth P. (2017). Na/K pump and beyond: Na/K-ATPase as a modulator of apoptosis and autophagy. Molecules, 22(4), 578. https://doi.org/10.3390/molecules22040578.

Ion channels are membrane-spanning proteins that create pores through which ions can passively diffuse across the membrane, following electrochemical gradients. These channels are extremely selective, allowing only certain ions such as sodium (Na^+), potassium (K^+), calcium (Ca^{2+}), and chloride (Cl^-) to pass through while blocking others. Ion channels are classified into three types based on their gating mechanisms: voltage-gated, ligand-gated, and mechanically gated ion channels (Gadsby, 2009).

Voltage-gated ion channels open and close in response to membrane potential changes. These channels are critical for producing and propagating action potentials in neurons and muscle cells. For example, voltage-gated sodium channels (Na^+ channels) open quickly in response to membrane depolarization, allowing an influx of Na^+ ions to further depolarize the membrane. This depolarization causes voltage-gated potassium channels (K^+ channels) to open, allowing K+ ions to exit the cell, repolarizing and eventually hyperpolarizing the membrane, restoring the resting membrane potential. The sequential opening and closing of Na^+ and K^+ channels are critical for the rapid conduction of electrical signals along nerves (Pollard et al., 2022).

Ligand-gated ion channels, or ionotropic receptors, open in response to the binding of a specific chemical ligand, such as a neurotransmitter. A well-known example is the nicotinic acetylcholine receptor, a ligand-gated cation channel that opens when acetylcholine is bound, resulting in an influx of Na^+ and Ca^{2+} and an efflux of K^+. This ion movement causes a local depolarization called an excitatory postsynaptic potential (EPSP), which can trigger an action potential in neurons (Alberts et al., 2015). These channels are essential for synaptic transmission and play a role in a variety of signaling pathways that control a wide range of physiological functions, including muscle contraction and sensory perception.

Mechanically gated ion channels open in response to mechanical forces on the cell membrane. These channels are necessary for sensory transduction processes like touch, hearing, and osmoregulation. Piezo channels, for example, are mechanically activated and contribute to the conversion of mechanical stimuli into electrical signals in various sensory neurons (Litan and Langhans, 2015). Their function is critical for processes such as tactile sensation and blood pressure regulation.

Ion channels allow for the passive movement of ions down electrochemical gradients, whereas ion pumps actively transport ions against concentration gradients using energy from ATP hydrolysis. These pumps are necessary to maintain ion gradients across the cell membrane, which are critical for cellular functions such as maintaining cell volume, resting membrane potential, and driving secondary active transport (Stein, 2012).

The sodium-potassium pump (Na^+/K^+ ATPase) is one of the best understood ion pumps. It is an electrogenic pump that actively transports three Na^+ ions out of the cell and two K^+ ions into the cell, despite their concentration gradients. This process uses one molecule of ATP per cycle, which is hydrolyzed to provide the required energy. The Na^+/K^+ ATPase is responsible for maintaining high intracellular K^+ and low intracellular Na^+ concentrations, which are required for resting membrane potential and action potential generation (Forrest et al., 2011). Furthermore, the gradients established by the Na^+/K^+ ATPase are used to drive secondary active transport processes, such as the sodium-glucose cotransporter (SGLT), which regulates glucose uptake in the intestine and kidney.

Another important ion pump is the calcium pump (Ca^{2+} ATPase), which actively transports Ca^+ ions from the cell or into the sarcoplasmic reticulum in muscle cells. This pump contributes to low intracellular Ca^{2+} concentrations, which are essential for cellular signaling, muscle contraction, and neurotransmitter release. The Ca^{2+} ATPase in the sarcoplasmic reticulum (SERCA) is especially important for muscle relaxation because it sequesters Ca^{2+} back into the sarcoplasmic reticulum after a contraction, allowing the muscle to relax (Baynes and Dominiczak, 2009).

The hydrogen-potassium pump (H^+/K^+ ATPase) is another type of ion pump that is essential for maintaining the stomach's acidic environment. This pump actively transports H^+ ions into the stomach lumen in exchange for K^+ ions, utilizing energy from ATP hydrolysis. The H^+/K^+ ATPase secretes H^+, which is necessary for food digestion and pathogen killing (Litan and Langhans, 2015).

The chloride-bicarbonate exchanger (also known as the anion exchanger or AE1) is a transport protein that functions as an antiporter. It facilitates the exchange of chloride ions (Cl^-) and bicarbonate ions (HCO_3^+) across the plasma membrane, which is crucial for maintaining acid-base balance in the blood. In red blood cells, this exchanger plays a vital role in transporting carbon dioxide (CO_2) from tissues to the lungs by enabling the conversion of CO_2 to bicarbonate for transport in the plasma (Alberts et al., 2015; Pollard et al., 2022).

9.3. ENDOCYTOSIS AND EXOCYTOSIS

Endocytosis and exocytosis are important cellular processes that transport large molecules, particles, and even whole cells across the plasma membrane. These processes are essential for maintaining cellular homeostasis, regulating membrane composition, and allowing intercellular communication. Endocytosis and exocytosis are dynamic and highly regulated mechanisms in cell biology that rely on a complex interplay of proteins and lipids to orchestrate the trafficking of materials into and out of the cell (Alberts et al., 2015) (Figure 9.2).

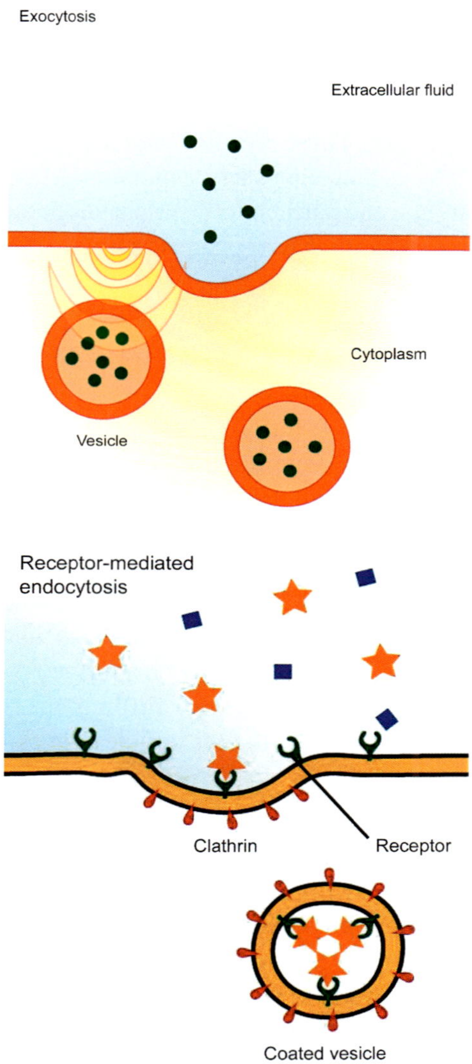

Figure 9.2. *Illustration of exocytosis and endocytosis.*

Source: https://courses.lumenlearning.com/suny-wmopen-biology1/chapter/endocytosis-and-exocytosis/.

Endocytosis is the process by which cells internalize extracellular substances by engulfing them in vesicles formed by the plasma membrane. This process is broadly classified into three types: phagocytosis, pinocytosis, and receptor-mediated endocytosis. Each type of endocytosis uses distinct mechanisms and serves specific physiological functions (Wu et al., 2014).

Phagocytosis, also known as "cellular eating," is a type of endocytosis where cells engulf larger particles like bacteria, dead cells, or debris. Phagocytic cells, such as macrophages and neutrophils, are primarily responsible for this process. Phagocytosis begins when the target particle is recognized and bound to specific receptors on the cell's surface. These receptors can identify pathogen-associated molecular patterns (PAMPs) and other surface markers. After binding, the plasma membrane expands around the particle to form pseudopodia, which eventually fuse to form a large vesicle known as a phagosome. The phagosome then fuses with lysosomes, forming a phagolysosome, where hydrolytic enzymes degrade the engulfed material (Alberts et al., 2015). This process is critical for immune defense, tissue remodeling, and the clearance of apoptotic cells.

Pinocytosis, also known as "cellular drinking," is a non-specific type of endocytosis in which extracellular fluids and dissolved solutes are taken up into small vesicles. This process is ongoing in most cell types and is responsible for the majority of fluid and nutrient internalization. Pinocytosis, unlike phagocytosis, does not involve any specific receptor-ligand interactions. Instead, it is caused by spontaneous invagination of the plasma membrane, which results in the formation of small vesicles known as pinocytic vesicles, which are then internalized. Pinocytosis is further classified into macropinocytosis and micropinocytosis based on the size of the vesicles formed (Oh and Park, 2014). Macropinocytosis creates large vesicles by ruffling the plasma membrane, whereas micropinocytosis produces smaller vesicles via clathrin-coated pits or caveolae.

Receptor-mediated endocytosis is a highly specific and efficient form of endocytosis that allows cells to selectively internalize specific molecules like hormones, nutrients, and growth factors. This process involves the binding of ligands to specific cell surface receptors, which causes plasma membrane invagination and the formation of clathrin-coated vesicles. These vesicles are coated with clathrin, a protein that is essential for shaping the vesicle and selecting cargo. When the vesicle is internalized, the clathrin coat sheds, and the uncoated vesicle fuses with early endosomes, where the acidic environment frequently causes the ligand-receptor complex to dissociate. The receptors can be recycled back to the plasma membrane, while the ligands are transported to lysosomes for degradation or to the trans-Golgi network for further processing (Oh and Park, 2014; Wu et al., 2014). Receptor-mediated endocytosis is essential for many cellular functions, such as nutrient uptake, receptor downregulation, and signal transduction.

Exocytosis, as opposed to endocytosis, is the process by which cells expel materials from their interior into the extracellular space. This process is necessary for the secretion of proteins, lipids, and other molecules, as well as the elimination of waste products. Exocytosis is broadly classified into two types: constitutive exocytosis and regulated exocytosis (Alberts et al., 2015).

Constitutive exocytosis occurs in all cells and is responsible for the continuous delivery of membrane proteins, lipids, and secretory proteins to the plasma membrane. This process is required to maintain membrane integrity, promote cell growth, and replenish membrane components lost during endocytosis. In constitutive exocytosis, cargo-containing vesicles emerge from the trans-Golgi network and travel to the plasma membrane, where they fuse and release their contents into the extracellular space. This process requires no specific signals and occurs in a relatively unregulated manner (Alberts et al., 2015; Pollard et al., 2022).

Regulated exocytosis, on the other hand, is a tightly controlled process that responds to specific signals, such as an increase in intracellular calcium levels. This type of exocytosis is especially common in specialized cells like neurons, endocrine cells, and immune cells, where it is required for the release of neurotransmitters, hormones, and other signaling molecules. During regulated exocytosis, vesicles containing cargo are stored in the cytoplasm near the plasma membrane until an appropriate signal causes them to fuse with the membrane. The signal, which frequently takes the form of an increase in intracellular Ca^{2+} concentration, activates many proteins, including the SNARE complex, which facilitates vesicle docking and fusion with the plasma membrane (Boal, 2012). This precisely timed release of vesicle contents is essential for processes such as synaptic transmission, insulin secretion, and the release of inflammatory mediators.

A complex network of proteins and lipids tightly regulates both endocytosis and exocytosis, ensuring vesicle trafficking specificity, timing, and fidelity. Small GTPases from the Rab and Arf families, for example, play critical roles in vesicle formation, movement, and fusion by serving as molecular switches that coordinate the assembly of protein complexes on vesicle membranes. Furthermore, phosphoinositides, a type of phospholipid, are involved in defining membrane identity and recruiting specific proteins to different membrane compartments, thus regulating vesicle trafficking and membrane dynamics (Wu et al., 2014).

The interaction between endocytosis and exocytosis is critical for maintaining cellular homeostasis and regulating various cellular processes. For example, the balance of these two processes determines plasma membrane composition and receptor surface expression, which affects cell signaling, nutrient uptake, and intercellular communication, respectively. Endocytosis and exocytosis dysfunction have been linked to a variety of diseases, including neurodegenerative disorders, cancer, and infectious diseases. Defects in the endocytic trafficking of amyloid precursor protein, for example, have been linked to Alzheimer's disease pathogenesis, whereas altered exocytosis of matrix metalloproteinases can contribute to tumor invasion and metastasis (Wu et al., 2014; Pollard et al., 2022).

9.4. CELL ADHESION MOLECULES

Cell adhesion molecules (CAMs) are essential components of cellular membranes that mediate cell adhesion to each other as well as to the extracellular matrix. These molecules are essential for tissue structure organization and maintenance, as well as various cellular processes such as signaling, migration, and differentiation. Cadherins, integrins, selectins, and members of the immunoglobulin superfamily (IgSF) are the four major categories of CAMs. Each family has distinct structural features and adhesive properties that contribute to the specific biological functions (Blankenberg et al., 2001) (Figure 9.3).\

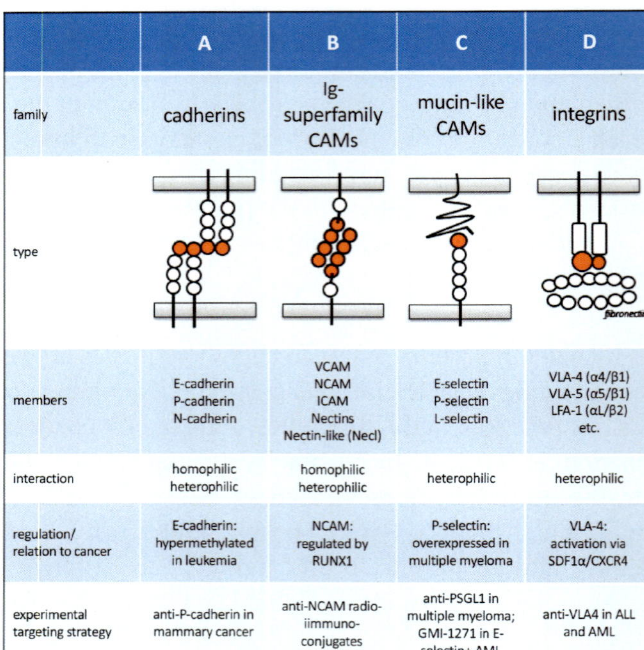

Figure 9.3. *Diagrammatic representation of four families of cell adhesion molecules (CAMS).*

Source: Windisch R, Pirschtat N, Kellner C, Chen-Wichmann L, Lausen J, Humpe A, Krause DS, Wichmann C. (2019). Oncogenic deregulation of cell adhesion molecules in leukemia. Cancers, 11(3):311. https://doi.org/10.3390/cancers11030311.

Cadherins are a class of calcium-dependent transmembrane proteins that mediate homophilic cell-cell adhesion, which means they bind to the same cadherin on adjacent cells. The extracellular domain of cadherins is made up of repeated cadherin domains that facilitate calcium binding, which is required for their adhesive function. Calcium ions stabilize the cadherin structure, preventing it from becoming flexible and ineffective. E-cadherin, a prototypical member of the cadherin family, is primarily expressed in epithelial tissues and is critical for maintaining epithelial cell integrity and barrier function. Cadherin-mediated adhesion is a dynamic process controlled by the actin cytoskeleton and intracellular signaling pathways. The intracellular domain of cadherins interacts with catenins, cytoplasmic proteins that connect cadherins to the actin cytoskeleton. This interaction is crucial for the strength and stability of cell-cell adhesion and for the transduction of signals that regulate cell behavior, such as proliferation and migration (Marshall et al., 2003).

Integrins are another major type of CAM that act as transmembrane receptors to mediate cell-ECM interactions. Integrins, unlike cadherins, are heterodimeric proteins made up of α and β subunits that regulate ligand binding and signaling. In humans, there are at least 18 α and 8 β subunits that combine to form 24 different integrin receptors, each with unique binding affinities and tissue distributions. Integrins are essential for transmitting mechanical and chemical signals between the extracellular matrix and the cell interior, a process known as "outside-in" and "inside-out" signaling. In "outside-in" signaling, integrin binding to ECM proteins such as fibronectin, collagen, and laminin activates intracellular signaling cascades that control a variety of cellular processes, including cell survival, proliferation, and migration. In contrast, "inside-out" signaling involves the activation of integrins by intracellular signals, which alter their conformation and affinity for ECM ligands, thereby modulating cell adhesion and motility (Pollard et al., 2022).

The selectin family of CAMs includes three members: L-selectin, E-selectin, and P-selectin, each named after the cell type in which they were first discovered (leukocytes, endothelium, and platelets, respectively). Selectins are carbohydrate-binding proteins that recognize specific glycosylated ligands on other cells' surfaces. This interaction promotes a transient, low-affinity type of adhesion that is required for leukocyte trafficking during immune surveillance and inflammation. For example, during an inflammatory response, endothelial cells increase the expression of E-selectin and P-selectin, which bind to sialyl Lewis X (sLex) glycoproteins on circulating leukocytes. This binding facilitates the initial "rolling" of leukocytes along the vascular endothelium, which is necessary for their firm adhesion and subsequent transmigration into inflamed tissues (Van Roy and Berx, 2008).

The immunoglobulin superfamily (IgSF) of CAMs is made up of proteins that have structural similarities to immunoglobulins and can mediate both homophilic and heterophilic adhesion. The IgSF includes neural cell adhesion molecules (NCAMs), intercellular adhesion molecules (ICAMs), and vascular cell adhesion molecules (VCAMs), among others. These CAMs are involved in a variety of biological processes, including immune cell trafficking, neuronal development, and tissue repair. NCAMs, for example, play an important role in neural development by mediating homophilic adhesion between neurons, allowing for axon guidance and synapse formation (Blankenberg et al., 2001). ICAM-1 and VCAM-1, on the other hand, are predominantly expressed on endothelial cells and are activated in response to inflammatory cytokines. These molecules bind to integrins on leukocytes, promoting firm adhesion and transmigration across the endothelium during immune responses.

CAMs serve more than just cell adhesion; they also play an important role in regulating intracellular signaling pathways that control cell behavior. CAM-mediated adhesive interactions can activate a variety of signal transduction cascades that affect gene expression, cytoskeletal dynamics, and cell fate decisions. Integrin engagement with ECM components, for example, can activate focal adhesion kinase (FAK) and Src family kinases, resulting in the phosphorylation of downstream targets that regulate cell survival, proliferation, and motility (Marshall et al., 2003). Similarly, cadherin-mediated cell-cell adhesion can activate the Hippo signaling pathway, which regulates transcription factors such as YAP and TAZ to control cell proliferation and apoptosis.

Furthermore, the function of CAMs is tightly controlled by a variety of mechanisms, including post-translational modifications, proteolytic cleavage, and intracellular traffic. Integrin endocytosis and recycling, for example, are critical for regulating cell adhesion and migration dynamically. Endocytosed integrins can either be recycled back to the plasma membrane or degraded, depending on the cellular context and signaling cues. CAM cleavage by extracellular proteases, such as matrix metalloproteinases (MMPs), is another regulatory mechanism that influences cell adhesion and signaling. For example, E-cadherin proteolytic shedding can disrupt cell-cell adhesion and promote epithelial-mesenchymal transition (EMT), a process linked to cancer progression and metastasis (Van Roy and Berx, 2008; Pollard et al., 2022).

CAM dysregulation has been linked to a variety of pathological conditions, including cancer, inflammation, and autoimmune disorders. In cancer, changes in CAM expression and function can contribute to tumor initiation, progression, and metastasis. The loss of E-cadherin expression, which is a hallmark of EMT, is common in invasive and metastatic cancers and is associated with a poor outcome. Similarly, abnormal integrin signaling can increase tumor cell survival, invasion, and resistance to treatment. Overexpression of selectins and IgSF CAMs can exacerbate leukocyte recruitment and tissue damage in inflammatory and autoimmune disorders. For example, the upregulation of VCAM-1 and ICAM-1 on endothelial cells is a critical event in the pathogenesis of atherosclerosis, rheumatoid arthritis, and multiple sclerosis (Blankenberg et al., 2001; Marshall et al., 2003).

9.5. CELL SIGNALING RECEPTORS

Cell signaling receptors are essential components of cellular communication that help maintain homeostasis, regulate cellular activities, and respond to environmental stimuli. These receptors are often proteins or glycoproteins found on the cell membrane, in the cytoplasm, or in the nucleus, and are designed to receive and send signals from a variety of extracellular substances such as hormones, neurotransmitters, and growth factors. When ligands connect to these receptors, they undergo conformational changes, triggering a cascade of intracellular events that eventually lead to specific physiological responses. Understanding the complexities of cell signaling receptors is critical in the science of cell biochemistry because it provides insights into the molecular mechanisms underpinning cellular function, disease development, and possible therapeutic targets.

The range of cell signaling receptors reflects the complexities of cellular communication. G-protein-coupled receptors (GPCRs), receptor tyrosine kinases (RTKs), and ion channel receptors are among the most extensively investigated types of these receptors. GPCRs, for example, have seven transmembrane domains and can activate intracellular G-proteins in response to ligand binding. This activation activates several downstream signaling pathways, including the cyclic AMP (cAMP) pathway, which regulates metabolism, cell development, and gene expression. RTKs, on the other hand, function by dimerizing and autophosphorylating upon ligand binding, activating numerous signaling cascades such as the Ras-MAPK pathway, which is critical for cell proliferation and differentiation (Chhichholiya et al., 2024). Meanwhile, ion channel receptors control the flow of ions across the cell membrane, which is critical for neuronal communication and muscle contraction.

The specificity and sensitivity of cell signaling receptors are critical for accurately regulating cellular responses. These receptors must be able to discriminate between different ligands and transmit the correct signals to trigger suitable cellular responses. This specificity is frequently achieved by structural complementarity of the ligand and the receptor's binding site, as well as the receptor's capacity to interact with certain intracellular signaling molecules.

For example, GPCRs' high ligand specificity ensures that diverse signals are generated in response to different external stimuli. These receptors are also highly sensitive, allowing cells to detect and respond to signaling molecules at extremely low concentrations. This sensitivity is frequently amplified by mechanisms such as receptor amplification, in which a single ligand-receptor interaction activates numerous downstream signaling molecules, magnifying the signal within the cell (Boal, 2012).

Cell signaling receptors are not only required for regular cellular function but they are also involved in a variety of disorders, including cancer. Dysregulation of receptor signaling, whether by mutations, overexpression, or abnormal activation, can result in uncontrolled cell proliferation, survival, and metastasis. Mutations in RTKs, such as the epidermal growth factor receptor (EGFR), have been linked to the development of several malignancies, including non-small cell lung cancer (NSCLC). In these instances, the receptor's persistent activation, even in the absence of a ligand, promotes cancer cell growth. Targeting these dysregulated receptors with particular inhibitors has become a cornerstone of cancer treatment. The employing of tyrosine kinase inhibitors (TKIs) to suppress the activity of mutant or overactive RTKs is an effective method in the treatment of malignancies such as NSCLC, demonstrating the therapeutic potential of modifying receptor signaling pathways (Boal, 2012).

Apart from their role in illness, cell signaling receptors are essential for the growth and operation of the immune system. The innate immune response depends on a family of pattern recognition receptors called toll-like receptors (TLRs). These receptors identify pathogen-associated molecular patterns (PAMPs) and start signaling cascades that trigger the activation of adaptive immune responses and cytokine production. The host's capacity to develop a successful immune response to pathogens depends on TLRs operating properly. On the other hand, abnormal TLR signaling activation can lead to autoimmune and inflammatory disorders, highlighting the fine balance in receptor signaling needed to preserve immunological homeostasis (Pollard et al., 2022; Chhichholiya et al., 2024). The study of cell signaling receptors remains a dynamic and fast-expanding area, with continuous research aiming at revealing the detailed processes of receptor function, signaling specificity, and involvement in health and illness. Techniques such as cryo-electron microscopy and single-cell RNA sequencing have revealed new insights into receptor shape and function, allowing for more targeted and successful therapeutic approaches. The capacity to control receptor activity, whether via small chemical inhibitors, monoclonal antibodies, or gene editing technologies, holds great promise for the treatment of a variety of disorders, emphasizing the relevance of cell signaling receptors in cell biochemistry.

9.6. CELL SIGNALING PATHWAYS

Cell signaling pathways are key frameworks that regulate and coordinate cellular functions in the field of cell biochemistry. These pathways form a sophisticated network of communication that cells employ to respond to internal and external stimuli, enabling a wide range of cellular operations such as growth, differentiation, metabolism, and death. Understanding the molecular mechanisms of cell signaling pathways is critical for understanding how cells maintain homeostasis and how disruption of these networks can lead to illnesses like cancer, diabetes, and neurodegenerative disorders.

Signaling substances such as hormones, neurotransmitters, cytokines, and growth factors are crucial to cell signaling pathways. These molecules bind to specific receptors on the cell surface or within the cell, causing a series of metabolic reactions. Receptor-ligand interactions start the signal transduction process, which frequently involves the activation of intracellular signaling proteins like kinases, phosphatases, and second messengers (Newton et al., 2016). The mitogen-activated protein kinase (MAPK) pathway is one of the best-studied signaling pathways, and it is critical for controlling cell proliferation, differentiation, and survival. When activated by extracellular signals, MAPKs phosphorylate target proteins, altering gene expression and cellular activity (Alberts et al., 2015).

Another important cell signaling system is the phosphoinositide 3-kinase (PI3K)/Akt pathway, which controls cell survival and metabolism. This route is initiated by growth hormones like insulin and results in the activation of Akt, a serine/threonine kinase. Akt then phosphorylates many downstream substrates involved in glucose metabolism, protein synthesis, and apoptosis suppression. Dysregulation of the PI3K/Akt pathway has been linked to a variety of malignancies, resulting in uncontrolled cell proliferation and survival (Nusse and Clevers, 2017).

The NF-κB pathway, which is important in immunological response, inflammation, and cell viability, is an additional example of a crucial signaling system. Genes involved in immunological and inflammatory responses are regulated by the transcription factor NF-κB, which translocates to the nucleus when activated. Pathogens, stress signals, and cytokines are some of the stimuli that tightly regulate the activation of the NF-κB pathway. Chronic inflammatory illnesses, autoimmune disorders, and cancer are linked to aberrant NF-κB signaling (Boutros et al., 2008). Cell signaling pathways are frequently linked together to form intricate webs of feedback loops and cross-talk that guarantee the exact regulation of cellular reactions. For example, maintaining a balance between apoptosis and cell growth requires the interaction of the PI3K/Akt and MAPK pathways. Maintaining tissue homeostasis and averting pathological diseases like cancer depend on this interaction. The function of scaffold proteins, which arrange signaling components into certain complexes to maintain signal specificity and fidelity, further highlights the intricacy of these networks (Newton et al., 2016).

Moreover, cell signaling involves the modification of lipids and other small molecules in addition to protein-protein interactions. For instance, the sphingolipid signaling pathway is essential for controlling choices about the destiny of cells, such as apoptosis, autophagy, and senescence. Bioactive lipid mediator sphingosine-1-phosphate (S1P) interacts with its receptors to initiate signaling pathways downstream that support the migration and survival of cells. Numerous clinical problems, such as cancer, cardiovascular diseases, and neurological disorders, have been associated with disruptions in sphingolipid metabolism (Alberts et al., 2015). In addition to biochemical characteristics, cell signaling pathways are regulated by the cell's spatial and temporal dynamics. The subcellular localization of signaling molecules, the timing of signal initiation and termination, and the compartmentalization of signaling complexes all influence the specificity and outcome of a signaling event. For example, Akt's localization in the plasma membrane is critical for its activation, although its translocation to other cellular compartments can influence its activity and downstream effects (Boutros et al., 2008). Similarly, the temporal dynamics of calcium signaling, characterized by oscillations and waves, have a substantial impact on the nature of the cellular response, which ranges from gene transcription to cell movement (Gomperts et al., 2009).

Technological developments have greatly improved our understanding of cell signaling networks. High-throughput techniques, including mass spectrometry-based proteomics and next-generation sequencing, have enabled the full mapping of signaling networks and the discovery of novel signaling components. Furthermore, the development of live-cell imaging and single-molecule tracking techniques has provided insights into the real-time dynamics of signaling events, allowing researchers to visualize the spatial and temporal aspects of signaling at unprecedented resolution (Newton et al., 2016).

Despite progress in understanding the molecular principles of cell signaling, numerous hurdles remain. The sheer complexity and redundancy of signaling networks, together with the variability of cellular responses, make it challenging to fully comprehend how these pathways operate in various biological situations. Furthermore, the integration of signaling pathways with other cellular functions, including metabolism and gene expression, adds a level of complexity that necessitates further exploration. Understanding these complex networks is critical for designing targeted therapeutics for diseases characterized by abnormal signaling, such as cancer and inflammatory disorders.

9.7. ROLE OF MEMBRANE TRANSPORT IN DISEASES

Membrane transport is a critical component of cellular function, regulating cellular homeostasis, signaling, and energy production. Its role extends beyond cellular upkeep to the development and progression of many illnesses. Membrane transport systems, which include passive and active transport, endocytosis, and exocytosis, are essential for regulating the passage of ions, nutrients, and other molecules across cellular membranes. Disruptions in these mechanisms can result in pathological situations, emphasizing the need to understand membrane transport in the context of disease.

Cystic fibrosis (CF) is one of the best-known examples of membrane transport failure that leads to illness. CF is caused by mutations in the cystic fibrosis transmembrane conductance regulator (CFTR) gene, which produces a chloride ion channel that regulates salt and water flow across epithelial membranes. Mutations in CFTR cause faulty chloride ion transport, resulting in thicker mucus discharges in the lungs, pancreas, and other organs. Chronic respiratory infections and pancreatic insufficiency, which are characteristics of cystic fibrosis, are caused by mucus accumulation. Understanding CFTR activity and its involvement in membrane transport has helped to design medicines that try to correct underlying transport abnormalities, such as CFTR modulators (Deretic et al., 2013).

Another notable example is the involvement of sodium-potassium ATPase (Na^+/K^+ ATPase) in cardiovascular disorders. Na^+/K^+ ATPase is a membrane-bound enzyme that maintains electrochemical gradients of sodium and potassium ions across the plasma membrane, which is necessary for many cellular activities such as nerve impulse transmission and muscle contraction. Dysregulation of Na^+/K^+ ATPase activity has been linked to hypertension and heart failure. Endogenous cardiotonic steroids, for example, block Na^+/K^+ ATPase, resulting in higher intracellular sodium levels that disrupt calcium homeostasis and stimulate vascular smooth muscle contraction, contributing to elevated blood pressure (Jain et al., 2021). Thus, Na^+/K^+ ATPase plays a critical role in the pathogenesis of cardiovascular disorders.

Membrane transport plays an important role in neurodegenerative illnesses such as Alzheimer's disease (AD). The amyloid precursor protein (APP) and its cleavage by membrane-bound enzymes play a critical role in Alzheimer's disease pathology. Improper APP processing by beta-secretase and gamma-secretase causes the accumulation of amyloid-beta (Aβ) peptides, which combine to form plaques, a characteristic of Alzheimer's disease (Jin et al., 2016). Membrane transport pathways play a role in the trafficking and localization of APP and the enzymes that digest it. Disruptions in these transport routes can increase the formation and deposition of Aβ in the brain, leading to neurodegeneration. Furthermore, disruption of ion transport in neurons, notably calcium and potassium, has been linked to synaptic dysfunction and neuronal death, aggravating the course of AD (Pollard et al., 2024).

Cancer is another illness that relies heavily on membrane transport systems. Overexpression of ATP-binding cassette (ABC) transporters, such as P-glycoprotein (P-gp), is frequently responsible for multidrug resistance (MDR) in cancer cells. These transporters aggressively efflux chemotherapeutic medicines from cancer cells, lowering intracellular concentrations and thereby limiting treatment efficacy (Gagneux et al., 2022). The significance of membrane transport in drug resistance has important implications for cancer treatment, necessitating the development of techniques to reduce or circumvent these transporters' actions. Furthermore, changes in ion transport, namely potassium and calcium channels, have been linked to cancer cell proliferation, migration, and metastasis, emphasizing the importance of membrane transport in cancer biology.

Another illness that relies heavily on membrane transport is diabetes mellitus, particularly type 2 diabetes. The insulin-mediated absorption of glucose into cells, which is predominantly assisted by glucose transporters (GLUTs), is an important mechanism in maintaining blood glucose levels. Insulin resistance affects GLUT4 translocation to the plasma membrane, resulting in decreased glucose absorption and hyperglycemia. This disruption of glucose transport contributes to the metabolic abnormalities seen in diabetes and associated consequences, including cardiovascular disease and neuropathy (Nusse and Clevers, 2017).

Hereditary hemochromatosis and Wilson's disease are two inherited disorders that are related to membrane transport. Wilson's illness is caused by mutations in the ATP7B gene, which codes for an ATPase that transports copper. These mutations impede the liver's ability to excrete copper, which accumulates in organs and causes damage to the liver and nervous system. Mutations in the HFE gene cause hereditary hemochromatosis, which results in excessive iron absorption and accumulation, mostly in the pancreas, liver, and heart, damaging and malfunctioning these organs (Brissot et al., 2018). These illnesses demonstrate the effects of membrane transport dysregulation and its vital role in metal homeostasis (Gagneux et al., 2022).

Chapter 10

Cell Biochemistry Techniques

Contents

Cell biochemistry encompasses a wide range of techniques used to understand cellular processes and their applications in real-world situations, particularly in medical and biotechnological fields. Western blotting, polymerase chain reaction (PCR), and mass spectrometry are important techniques for analyzing proteins and nucleic acids, allowing researchers to identify biomarkers for diseases like cancer and genetic disorders. For example, Western blotting allows for the detection of specific proteins within a complex mixture, providing insights into cellular signaling pathways and their alterations in disease states (Joo et al., 2008).

Similarly, PCR amplifies DNA sequences, which aids in the diagnosis of infectious diseases by detecting pathogen-specific DNA. Furthermore, mass spectrometry provides high-throughput identification and quantification of proteins and metabolites, which is critical for understanding cellular metabolism and its dysregulation in diseases such as diabetes. These techniques also support advances in personalized medicine, which uses patient-specific molecular profiles to guide targeted therapies. In agriculture, cell biochemistry techniques are used to improve crop resistance and productivity through genetic modification and stress response analysis. Together, these biochemistry techniques provide a foundational framework for transforming cellular insights into practical applications across diverse scientific and clinical disciplines (Davis, 2012).

10.1. CELL FRACTIONATION AND ISOLATION

Cell fractionation and isolation are fundamental techniques in cell biochemistry that enable researchers to unravel the complexities of cellular architecture and comprehend the functions of individual components. These methodologies are critical for elucidating cellular processes, studying biochemical pathways specific to various organelles, and investigating disease mechanisms at the molecular level (Pollard et al., 2022) (Figure 10.1).

Cell fractionation is the process of separating cells into their constituent organelles and macromolecules. This process typically begins with cell disruption, which can be accomplished mechanically, chemically, or enzymatically. Mechanical disruption methods, such as homogenization, sonication, and grinding, physically break down cells through force, whereas chemical and enzymatic methods use detergents or enzymes to dissolve cell membranes. The method used is determined by the cell type and the subcellular structures of interest, to maintain the integrity and functionality of the components being isolated (Wyatt Shields Iv et al., 2015).

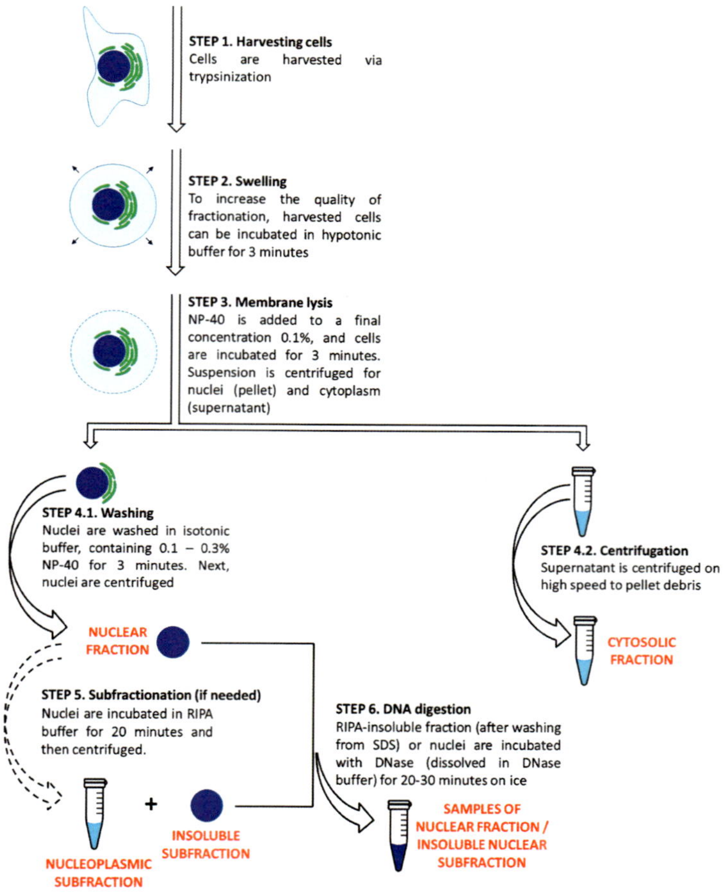

Figure 10.1. *Graphical representation of the nucleus/cytoplasm fractionation.*

Source: Senichkin VV, Prokhorova EA, Zhivotovsky B, Kopeina GS. (2021). Simple and efficient protocol for subcellular fractionation of normal and apoptotic cells. Cells, 10(4):852. https://doi.org/10.3390/cells10040852.

When cells are disrupted, the resulting homogenate contains a variety of cellular components that must be separated based on their physical and biochemical properties. Differential centrifugation is the most commonly used technique for this separation. This method involves spinning the homogenate at different speeds, allowing organelles to settle based on their size and density. Low-speed centrifugation removes large debris and unbroken cells first, then medium-speed centrifugation pellets nuclei and other large organelles, and high-speed or ultracentrifugation separates smaller organelles such as mitochondria, lysosomes, and peroxisomes. By gradually increasing the centrifugation speed, researchers can obtain relatively pure fractions of different organelles (Abubakar and Haque, 2020).

Beyond differential centrifugation, density gradient centrifugation provides a more refined method for separating organelles. This method uses a gradient of a dense substance, such as sucrose or cesium chloride, inside the centrifuge tube. Organelles migrate through the gradient and settle at locations where their density is equal to that of the surrounding medium. This technique is especially useful for isolating organelles of similar size but different densities, such as lysosomes and mitochondria. Density gradient centrifugation can also be tailored to isolate macromolecular complexes and membrane fractions, providing further insights into the composition and function of these cellular structures (Baynes and Dominiczak, 2009).

Isolating specific organelles is an important step in understanding their distinct biochemical properties and functions. For example, isolating mitochondria allows researchers to investigate their role in energy production and apoptosis. Scientists have studied isolated mitochondria to understand the mechanisms of oxidative phosphorylation and to identify key components involved in mitochondrial respiration. Similarly, isolating lysosomes has provided insights into cellular degradation processes and revealed how lysosomal dysfunction can cause diseases such as lysosomal storage disorders (Abubakar and Haque, 2020).

Modern advances in cell fractionation techniques have greatly improved our ability to study cellular functions in a very specific manner. Immunoaffinity methods, for example, employ antibodies to selectively bind and isolate specific organelles or protein complexes from a cell lysate. These methods, often combined with magnetic beads or chromatography, provide high specificity and purity, allowing for detailed analysis of organelle-specific processes. Immunoaffinity purification has helped identify protein complexes involved in cellular signaling pathways and understand their regulation (Wyatt Shields Iv et al., 2015).

Furthermore, cell isolation techniques have evolved to allow for the study of individual cells and organelles, providing previously unattainable insights into cellular heterogeneity. Individual cells can be isolated from a heterogeneous population using techniques such as flow cytometry and laser capture microdissection, which rely on specific markers or morphology. These isolated cells can then be subjected to further fractionation to study their organelle composition and functionality, offering a better understanding of cellular diversity and its implications for development, differentiation, and disease (Alberts et al., 2015). The use of cell fractionation and isolation surpasses basic research, with significant implications in clinical and biotechnological fields. In medicine, these techniques are used to identify disease biomarkers and create targeted therapies. For example, isolating cancer cell-derived exosomes from blood samples has emerged as a non-invasive method for cancer diagnosis and monitoring, providing useful information about tumor progression and treatment response (Wyatt Shields Iv et al., 2015). In biotechnology, cell fractionation is used to isolate specific enzymes or proteins for use in industrial applications such as drug development and biofuel production.

10.2. PROTEIN PURIFICATION METHODS

Protein purification is a basic technique in cell biochemistry that involves isolating a specific protein from a complex mixture, such as a cell lysate, in order to investigate its structure, function, and interactions. Protein purification is important for many applications, including enzyme characterization, drug development, and understanding the molecular mechanisms that underpin biological processes. Several methods, each based on distinct physicochemical properties of proteins, are used to achieve high purity and yield, with the method chosen depending on the target protein's characteristics and the intended application (Pollard et al., 2022) (Figure 10.2).

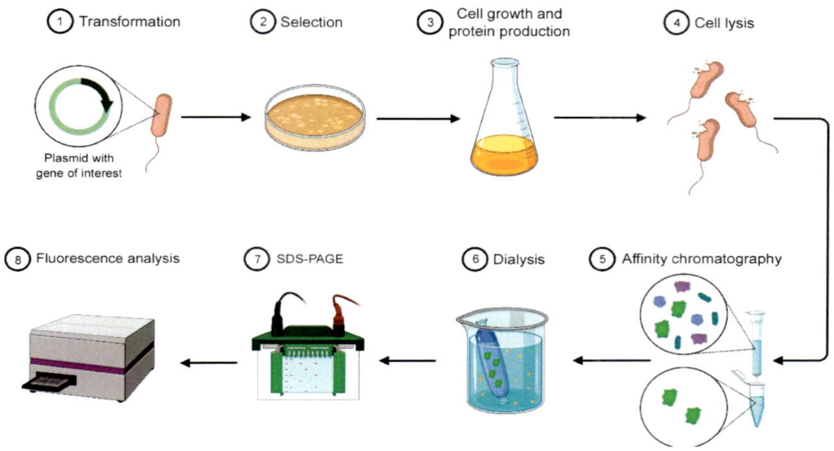

Figure 10.2. *Workflow of a typical protein purification procedure.*

Source: Gomes LC, Ferreira C, Mergulhão FJ. (2022). Implementation of a practical teaching course on protein engineering. Biology, 11(3):387. https://doi.org/10.3390/biology11030387.

Chromatography is a widely used method of protein purification that separates proteins based on their size, charge, hydrophobicity, or affinity for specific ligands. Size exclusion chromatography (SEC), also known as gel filtration chromatography, separates proteins according to their size and shape. In SEC, a column is filled with porous beads that form a matrix through which proteins pass. Larger proteins are excluded from the pores and elute first, whereas smaller proteins enter and elute second. This technique is particularly useful for purifying proteins that are sensitive to changes in pH or ionic strength, as it requires no alteration of the buffer composition (Davis, 2012).

Ion exchange chromatography (IEC) takes advantage of proteins' charge properties to facilitate separation. Proteins are first loaded into a column that contains charged resin. Proteins will bind to anion or cation exchange resins based on their net charge at a given pH level. Proteins are eluted by gradually increasing the salt concentration in the buffer, which competes with the proteins for binding to the resin's charged groups. Proteins with fewer interactions elute at lower salt concentrations, whereas proteins with more interactions elute at higher salt concentrations. This method allows for high-resolution separation of proteins with slight charge differences and is commonly used in the initial stages of purification (Janson, 2011).

Affinity chromatography is a potent technique that makes use of specific interactions between a protein and a ligand immobilized on a chromatography matrix. Nickel-affinity chromatography, for example, can be used to purify histidine-tagged proteins by binding histidine residues to nickel ions on the resin. The bound protein is then eluted with imidazole, which competes with the histidine residues for binding to the nickel ions. Affinity chromatography is the preferred method for recombinant protein purification due to its high specificity and ability to significantly enrich the target protein in a single step (Davis, 2012).

Another important technique is hydrophobic interaction chromatography (HIC), which separates proteins according to their hydrophobicity. Proteins are loaded onto a column under high salt conditions, which enhances the hydrophobic interactions between the protein and the resin's hydrophobic groups. As the salt concentration gradually decreases, proteins elute according to their hydrophobicity, with more hydrophobic proteins eluting later. HIC is especially useful for purifying proteins that are prone to aggregation, because the high salt concentrations used in the binding buffer can help maintain protein solubility (Alberts et al., 2015).

In addition to chromatography, centrifugation-based techniques are essential in protein purification. Differential centrifugation is a preparative technique for separating proteins based on their size and density. By spinning a cell lysate at increasing speeds, different cellular components, such as organelles and macromolecular complexes, can be separated. Proteins can be further purified from these fractions using density gradient centrifugation, which separates proteins according to their buoyant density in a gradient medium. These methods are especially useful for isolating membrane proteins and other proteins related to cellular organelles (Davis, 2012).

Electrophoretic techniques such as sodium dodecyl sulfate-polyacrylamide gel electrophoresis (SDS-PAGE) and isoelectric focusing (IEF) are also required for protein purification and analysis. SDS-PAGE separates proteins based on their molecular weight by denaturing them with SDS, which results in a uniform negative charge. Proteins are then resolved using a polyacrylamide gel, with smaller proteins migrating faster to the anode. This technique is widely used to determine the purity of protein preparations and to estimate molecular weights. In contrast, isoelectric focusing separates proteins according to their isoelectric points (pI). Proteins are subjected to an electric field in a pH gradient and migrate to the point where their net charge equals zero. IEF is particularly useful for resolving proteins with similar molecular weights but different pI values, making it an excellent complementary technique to SDS-PAGE (Baynes and Dominiczak, 2009).

Two-dimensional gel electrophoresis (2D-GE) is a technique that combines SDS-PAGE and IEF to separate complex protein mixtures at high resolution. Proteins are separated in the first dimension by pI and in the second dimension by molecular weight using IEF and SDS-PAGE, respectively. This method allows for the separation and analysis of thousands of proteins concurrently and is particularly useful in proteomics for identifying protein expression patterns and post-translational modifications (Davis, 2012).

Protein activity and function must be preserved during purification. When proteins are exposed to harsh conditions such as extreme pH, high salt concentrations, or denaturing agents, their biological activity can be reduced. To maintain protein stability and functionality during the purification process, it is critical to optimize purification conditions such as buffer composition, temperature, and the use of protease inhibitors (Janson, 2011).

The final step in protein purification frequently entails concentrating the purified protein and transferring it to a buffer suitable for downstream applications. Ultrafiltration and dialysis are popular methods for this purpose. Ultrafiltration employs semi-permeable membranes to retain proteins while allowing small molecules and salts to pass through, concentrating the protein solution. Dialysis, on the other hand, uses a semi-permeable membrane to exchange the buffer, removing unwanted small molecules and salts while preserving the protein of interest. Both methods are gentle and contribute to protein stability, making them ideal for preparing proteins for functional assays and structural studies (Davis, 2012; Janson, 2011).

10.3. DNA SEQUENCING AND ANALYSIS

DNA sequencing and analysis are critical techniques in cell biochemistry that have transformed our understanding of genetic information and its function in biological processes. These methodologies allow for the precise order of nucleotides within a DNA molecule, which is critical for discovering the genetic basis of diseases, understanding evolutionary relationships, and advancing fields such as genomics and biotechnology. Several DNA sequencing techniques have been developed over the years, each with its own principles, benefits, and applications (Alberts et al., 2015).

Frederick Sanger developed the Sanger sequencing technique in 1977, which is the earliest and most fundamental method of DNA sequencing. This method, also known as the chain-termination method, selectively incorporates chain-terminating dideoxynucleotides (ddNTPs) during DNA replication. In Sanger sequencing, the DNA to be sequenced is first denatured into single strands before being annealed to a short primer that serves as the starting point for DNA synthesis. A DNA polymerase enzyme extends the primer by adding nucleotides that are complementary to the template strand. However, the inclusion of ddNTPs, which lack a 3'-hydroxyl group necessary for forming a phosphodiester bond, causes the termination of DNA strand elongation upon incorporation (Davis, 2012).

The Sanger method was initially used to visualize DNA fragments with radiolabeled ddNTPs and autoradiography. However, advances in fluorescence detection, combined with capillary electrophoresis, have increased its throughput, accuracy, and safety, enabling sequencing automation and widespread use in genomic research. Despite its limitations in terms of sequencing speed and cost, Sanger sequencing remains a gold standard for its high accuracy and is widely used to validate the results obtained from high-throughput sequencing technologies (Pollard et al., 2022).

The advent of next-generation sequencing (NGS) technologies has significantly increased the speed and scale of DNA sequencing. NGS encompasses a variety of platforms that allow for the parallel sequencing of millions of DNA fragments, significantly reducing the time and cost of sequencing large genomes. One of the most popular NGS methods is sequencing by synthesis, which is implemented in the Illumina platform. This technique involves immobilizing fragmented DNA on a solid surface, such as a flow cell, and amplifying it in situ to form clusters of identical DNA molecules. Sequencing is then performed by sequentially adding labeled nucleotides and imaging the resulting incorporation events with high-resolution cameras. The fluorescent signals emitted by the labeled nucleotides are detected and recorded, providing real-time data about the DNA sequence (Baynes and Dominiczak, 2009).

The 454 Life Sciences platform pioneered pyrosequencing, another NGS approach. Pyrosequencing detects the release of pyrophosphate during nucleotide incorporation, which initiates a series of enzymatic reactions that produce light. The intensity of the emitted light is proportional to the number of nucleotides, making it possible to determine homopolymeric sequences. While pyrosequencing was one of the first NGS technologies to be commercialized and offered advantages in terms of read length, it has largely been replaced by other NGS methods due to limitations in sequencing accuracy and scalability (Davis, 2012).

Another NGS method, Ion Torrent sequencing, detects pH changes caused by the release of hydrogen ions during DNA synthesis. In this method, DNA fragments are attached to beads and amplified using emulsion PCR, generating millions of copies of each fragment. The beads are then placed in wells on a semiconductor chip, where nucleotides are sequentially flowed across the chip. As each nucleotide is incorporated, the release of hydrogen ions alters the pH, which is detected by the chip's sensors, allowing for real-time sequencing without the use of fluorescent labels. Ion Torrent sequencing provides a relatively fast and cost-effective alternative for targeted sequencing applications, but it may be less suitable for applications that require high accuracy (Davis, 2012).

Third-generation sequencing technologies, such as those developed by Pacific Biosciences (PacBio) and Oxford Nanopore Technologies, have increased the capabilities of DNA sequencing. PacBio's single-molecule real-time (SMRT) sequencing uses fluorescently labeled nucleotides to observe DNA synthesis in real-time. Unlike previous methods, SMRT sequencing does not require amplification, which reduces the possibility of introducing biases and errors. This method is especially useful for sequencing repetitive regions and complex genomes because it can generate long reads spanning tens of kilobases, providing more context for sequence assembly and structural variation analysis (Pollard et al., 2022).

Oxford Nanopore Technologies has created a nanopore-based sequencing method that also enables real-time analysis of individual DNA molecules. This method involves passing DNA molecules through a nanopore embedded in a membrane and measuring the changes in ionic current that occur as nucleotides pass through the pore. These changes are specific to the nucleotide sequence, allowing for direct and rapid sequencing of native DNA strands. Nanopore sequencing provides unparalleled read lengths and the ability to sequence various nucleic acid modifications, making it an effective tool for de novo genome assembly and epigenetic studies (Alberts et al., 2015).

DNA sequencing data analysis is an important part of sequencing workflows, requiring several computational steps to ensure accurate interpretation of raw sequence data. The first step in data analysis is base calling, which converts the raw signals generated during sequencing into nucleotide sequences. To ensure data reliability, quality control measures such as trimming low-quality bases and filtering out contaminant sequences are implemented. The cleaned reads are then aligned to a reference genome or assembled de novo, depending on the sequencing goals. For resequencing studies, alignment to a reference genome facilitates the identification of single nucleotide polymorphisms (SNPs), insertions, deletions, and structural variants, which are critical to understanding genetic diversity and disease associations (Baynes and Dominiczak, 2009).

De novo assembly involves computationally assembling reads into longer contiguous sequences, or contigs, based on sequence overlap. This procedure is especially important for sequencing projects involving novel genomes or highly variable genomic regions where reference sequences are either unavailable or incomplete. De novo assembly algorithms, such as De Bruijn graph-based methods, allow for efficient assembly of large genomes from short-read data by breaking the sequences into smaller k-Mers and reconstructing the original sequence based on k-mer overlaps (Davis, 2012).

Functional annotation, which involves identifying and characterizing genes and regulatory elements, may be used to further analyze sequencing data. This step entails comparing the assembled sequences to existing databases, predicting gene models, and deducing biological functions based on sequence homology. Integrative analysis, which combines DNA sequencing with other omics data like transcriptomics, proteomics, and epigenomics, can provide a more comprehensive view of gene regulation and cellular processes (Baynes and Dominiczak, 2009).

10.4. MASS SPECTROMETRY IN PROTEOMICS

Mass spectrometry (MS) is a powerful analytical technique used in proteomics to identify, characterize, and quantify proteins in complex biological samples. Proteomics, the large-scale study of proteins, their structures, functions, and interactions, heavily relies on MS to understand the proteome's complexity and provide insights into cellular biochemistry and molecular biology. The use of MS in proteomics has transformed our understanding of cellular processes by allowing detailed analysis of protein compositions, post-translational modifications (PTMs), and protein-protein interactions (Alberts et al., 2015) (Figure 10.3).

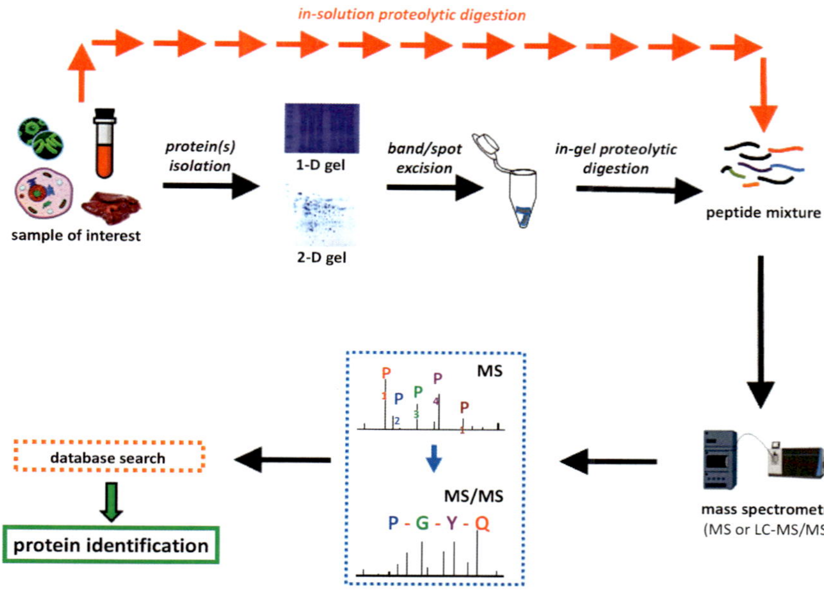

Figure 10.3. *Schematic overview of a bottom-up proteomics experiment.*

Source: Dupree EJ, Jayathirtha M, Yorkey H, Mihasan M, Petre BA, Darie CC. (2020). A critical review of bottom-up proteomics: the good, the bad, and the future of this field. Proteomes, 8(3):14. https://doi.org/10.3390/proteomes8030014.

The fundamental principle of mass spectrometry is to ionize chemical compounds to produce charged molecules or molecule fragments, then measure their mass-to-charge ratio (m/z) and detect these ions to obtain a mass spectrum. In proteomics, proteins of interest are first digested into smaller peptides, usually with an enzyme like trypsin, which cleaves proteins at specific sites to produce peptides suitable for MS analysis (Pollard et al., 2022). The peptides are then loaded into the mass spectrometer, where they are ionized, separated, and detected.

One of the most common ionization techniques used in proteomics is electrospray ionization (ESI), which generates ions from liquid samples. ESI involves passing the peptide solution through a fine needle under a high voltage, resulting in a fine mist of charged droplets. As the solvent evaporates, the charged peptides remain behind and enter the mass spectrometer in the gas phase. ESI is particularly suitable for coupling with liquid chromatography (LC), a technique that separates peptides based on their hydrophobicity, allowing high-throughput analysis of complex mixtures (Han et al., 2008).

Another popular ionization method is matrix-assisted laser desorption/ionization (MALDI). Peptides and a matrix compound are co-crystallized on a metal plate during MALDI. A laser pulse excites the matrix, causing the peptides to desorb and ionize into the gas phase. MALDI is commonly combined with time-of-flight (TOF) mass analyzers, which separate ions based on their flight time in a vacuum tube, allowing the analysis of large biomolecules with high accuracy and sensitivity (Baynes and Dominiczak, 2009).

Once ionized, the peptides are fed into a mass analyzer, which separates ions based on their m/z ratio. Proteomics makes use of a variety of mass analyzers, including quadrupole, ion trap, time-of-flight, and orbitrap. Each analyzer has distinct features that make it suitable for a variety of proteomic studies. For example, quadrupole and ion trap analyzers are frequently used for tandem mass spectrometry (MS/MS) experiments, where peptides are fragmented to provide sequence information, whereas orbitrap analyzers offer high resolution and mass accuracy, making them ideal for identifying PTMs and performing quantitative proteomics (Davis, 2012; Shuken, 2023).

MS/MS, also known as tandem mass spectrometry, is a critical step in proteomic analysis that allows for the identification of peptides based on their fragmentation patterns. In an MS/MS experiment, peptide ions (precursors) are fragmented in a collision cell, usually through collision-induced dissociation. The fragment ions are then analyzed in a second stage of mass spectrometry. The MS/MS spectra generated show a series of peaks corresponding to fragment ions, which can be used to determine the amino acid sequence of the peptide. This information is then matched against protein databases to determine the corresponding proteins (Shuken, 2023).

Quantitative proteomics, which seeks to determine the abundance of proteins in various biological states, is also heavily dependent on MS. Many quantitative MS strategies have been developed, including label-free quantification and stable isotope labeling. Label-free quantification compares peptide signal intensities across different samples, whereas stable isotope labeling involves incorporating isotopic labels into proteins or peptides, such as SILAC (stable isotope labeling by amino acids in cell culture), iTRAQ (isobaric tags for relative and absolute quantitation), or TMT (tandem mass tags), allowing for relative quantification in a single MS experiment. These approaches allow for comprehensive analysis of proteomic changes in response to various conditions, providing insights into cellular pathways and mechanisms (Han et al., 2008; Alberts et al., 2015).

Mass spectrometry can also be used to study PTMs, which are important for regulating protein function and cellular signaling. PTMs such as phosphorylation, acetylation, ubiquitination, and glycosylation can be identified and localized to specific amino acid residues using MS/MS. Advanced techniques have been developed to enhance the detection of PTMs, allowing for a more detailed understanding of their roles in cellular processes. These techniques include electron transfer dissociation (ETD) and higher-energy collisional dissociation (HCD) (Pollard et al., 2022).

MS's integration with other technologies, such as bioinformatics and systems biology, has broadened its applications in proteomics. Computational tools are required for analyzing the massive amounts of data produced by MS, such as protein identification, quantification, and pathway analysis. These tools allow for the reconstruction of protein networks as well as the identification of key regulatory proteins and pathways, which helps to discover new biomarkers and therapeutic targets (Han et al., 2008; Shuken, 2023).

10.5. FLUORESCENCE MICROSCOPY

Fluorescence microscopy has transformed cell biochemistry by providing an effective tool for visualizing molecular and cellular features. This technology relies on fluorescent molecules that release light when activated by specific wavelengths, allowing researchers to see biological components with high precision and clarity. Unlike traditional microscopy, which is based on light absorption and reflection, fluorescence microscopy uses the intrinsic features of fluorescent dyes, proteins, or other molecules to label and track specific biological targets. These fluorophores absorb light at one wavelength and emit it at a longer wavelength, enabling the selective imaging of certain structures within complicated biological settings (Wyatt Shields Iv et al., 2015).

Fluorescence microscopy has evolved in concert with advances in fluorescent probe technology. Early research relied on naturally occurring fluorescent molecules, such as green fluorescent protein (GFP), which was initially obtained from the jellyfish *Aequorea victoria*. GFP and its derivatives have now become essential tools in cell biochemistry because they can be genetically encoded and fused to a wide range of proteins to see their distribution and dynamics in living cells (Abubakar and Haque, 2020). The capacity to examine proteins in their natural environment has revealed previously unattainable insights into biological processes such as protein-protein interactions, signaling pathways, and compartmental architecture.

One of the most notable benefits of fluorescence microscopy is its ability to produce high contrast and specificity. Researchers can label several structures at the same time using fluorophores with varied excitation and emission spectra, allowing them to analyze interactions between various biological components. This multiplexing feature is especially useful in the investigation of complicated biological systems, where the behavior of numerous molecules must be monitored in real-time. Förster resonance energy transfer (FRET) and fluorescent lifetime imaging microscopy (FLIM) have been developed to make use of these features further, enabling nanometer-scale measurements of molecular distances and interactions (Janson, 2011).

Furthermore, the introduction of super-resolution fluorescence microscopy has pushed optical resolution beyond the diffraction limit, which previously limited light microscopy to a resolution of around 200 nanometers. Stimulated emission depletion (STED), photoactivated localization microscopy (PALM), and stochastic optical reconstruction microscopy (STORM) have all made it possible to visualize structures at the molecular size, down to tens of nanometers. These developments have had a transformative impact in domains such as neurobiology, where the ability to resolve synaptic structures and protein complexes at this scale has brought new insights into neural communication and plasticity (Wyatt Shields Iv et al., 2015).

Fluorescence microscopy is very important in the study of cellular dynamics. Time-lapse fluorescence microscopy, for example, enables the monitoring of biological processes over time, providing a dynamic perspective of activities such as cell division, migration, and intracellular trafficking. Live-cell imaging, made possible by breakthroughs in fluorescent protein technology and microscopy techniques, is now a basic tool in cell biology. It enables researchers to watch the real-time cellular behavior in reaction to stimuli, providing a better understanding of cellular function and control (Janson, 2011).

Fluorescence microscopy has become a cornerstone in biomedical sciences, particularly diagnostics and therapeutics, in addition to its uses in basic research. For example, fluorescence in situ hybridization (FISH) is a technique that employs fluorescent probes to identify and localize the presence or absence of certain DNA sequences on chromosomes. This approach is commonly utilized in the detection of genetic illnesses and malignancies, as it can identify chromosomal abnormalities that cause disease states. Similarly, fluorescence-based assays are widely employed in drug development to screen for molecules that modify biological targets, with high-throughput fluorescence microscopy allowing for the simultaneous analysis of thousands of samples (Abubakar and Haque, 2020).

Furthermore, the combination of fluorescence microscopy and other imaging modalities has increased its possibilities. By combining fluorescence with techniques like electron microscopy, researchers may link high-resolution structural data with fluorescence-based molecular information. This correlative light and electron microscopy (CLEM) technique provides a more complete understanding of cellular architecture and function, allowing for the investigation of subcellular structures with high resolution and molecular specificity (Davis, 2012). The fluorescence microscopy technique has several limitations. Photobleaching, or the permanent degradation of a fluorophore by light exposure, remains a serious difficulty, particularly in long-term imaging studies. Phototoxicity, or damage to living cells induced by prolonged exposure to bright light, is another issue, particularly in live-cell imaging. These concerns demand careful tuning of imaging parameters to balance high-quality data capture with sample integrity preservation (Pollard et al., 2022).

In recent years, the introduction of more photostable and less noxious fluorescent dyes, as well as developments in microscopy hardware, have helped to alleviate some of these restrictions. Light-sheet fluorescence microscopy (LSFM), which illuminates the sample with a thin sheet of light, has reduced photobleaching and phototoxicity by limiting the specimen's exposure to light. LSFM also enables quick imaging of large samples, making it ideal for three-dimensional imaging of entire organs or embryos (Davis, 2012).

The ongoing development of fluorescence microscopy techniques and technology emphasizes its critical significance in cell biochemistry and biomedical research. As researchers push the boundaries of what can be seen and quantified inside cells, fluorescence microscopy will definitely continue to yield crucial insights into the molecular systems that govern cellular function and illness.

10.6. WESTERN BLOTTING AND IMMUNOHISTOCHEMISTRY

Western blotting and immunohistochemistry are two essential cell biochemistry techniques that have greatly aided our understanding of protein expression, location, and interaction in biological systems. Both approaches are effective instruments for investigating biological processes and have proven crucial in many disciplines of research, including molecular biology, cancer, and neurology. While all methods have the same purpose of identifying specific proteins, they differ in their applications, methodologies, and information content, making each methodology distinct in the context of cell biochemistry.

Western blotting, or protein immunoblotting, is a popular analytical method for identifying and measuring certain proteins in a complicated mixture. Proteins are extracted from a sample first, usually by cell lysis. Next, proteins are separated based on molecular weight using sodium dodecyl sulfate-polyacrylamide gel electrophoresis (SDS-PAGE). After being separated, the proteins are placed onto a membrane, which is typically composed of PVDF or nitrocellulose, and are then probed with target protein-specific antibodies. Once the target protein has been bound by the primary antibody, the protein-antibody complex can be seen using a secondary antibody that has been coupled with either an enzyme or a fluorescent marker. The target protein is subsequently detected using techniques like chemiluminescence, fluorescence, or colorimetric assays; the strength of the signal indicates how abundant the protein is (Punekar, 2018).

Western blotting is especially useful since it may offer quantitative data on protein expression levels, making it an important technique for analyzing differential protein expression in a variety of situations, such as disease states or therapy responses. Western blotting, for example, is frequently employed in cancer research to measure the expression of oncogenes or tumor suppressor proteins, providing information about the molecular pathways that drive tumor progression (Pollard et al., 2024). Furthermore, Western blotting can be used to validate the specificity of antibodies employed in other techniques, such as immunohistochemistry, verifying that the observed staining patterns are truly caused by the target protein.

In contrast, immunohistochemistry (IHC) is a technique for visualizing protein localization within tissue sections or cell samples. Unlike Western blotting, which involves protein extraction and denaturation, IHC maintains the tissue's structural integrity, allowing researchers to investigate protein expression concerning cellular shape and tissue architecture. IHC is the procedure of applying antibodies to fixed tissue sections that have been processed to preserve antigens. Similar to Western blotting, a primary antibody binds to the target protein, while a secondary antibody coupled with a detection marker aids in visibility. However, in IHC, detection is frequently carried out using chromogenic substrates that create a colorimetric signal or fluorescent dyes that allow for the study of protein distribution under a microscope.

The use of IHC in histopathology is very essential since it is used to diagnose diseases based on the presence and distribution of specific biomarkers in tissue samples. For example, IHC is commonly used in cancer diagnostics to identify the expression of hormone receptors, such as estrogen and progesterone receptors in breast cancer, which helps guide treatment options. Given that IHC can offer spatial information regarding protein expression within tissues, it is an invaluable tool for examining cellular processes in situ, including differentiation, apoptosis, and cell signaling (Shuken, 2023).

Both Western blotting and immunohistochemistry rely on antibody specificity to detect target proteins, yet each has advantages and disadvantages. Western blotting is highly sensitive and can be used to quantify protein levels, but it lacks the spatial resolution that IHC provides. On the other hand, IHC provides precise information about protein localization within tissues, but it is generally less quantitative and can be affected by antibody binding efficiency and tissue processing. The study subject at hand frequently determines which technique to use. Western blotting may be recommended for studies that need exact protein level quantification, but IHC is more suited for studying protein distribution and localization within tissues (Han et al., 2008).

Furthermore, the combination of Western blotting and IHC can provide complementary information that improves our overall understanding of cellular processes. Western blotting, for example, can quantify a protein's overall expression in a tissue sample, whereas IHC can disclose its particular localization inside different cell types or tissue locations. This integrative method is also valuable in translational research, where understanding protein levels and distribution is essential for creating tailored therapeutics (Han et al., 2008).

Both approaches' capabilities have grown as a result of technological advances. Antibody development innovations, such as the synthesis of monoclonal antibodies, have enhanced the specificity and sensitivity of both Western blotting and IHC. Furthermore, multiplexing techniques in IHC enable the simultaneous detection of several proteins inside a single tissue segment, offering a more complete picture of protein interactions and cellular phenotypes (Pollard et al., 2024). Similarly, digital imaging and software analysis have improved the quantitative components of IHC, allowing for more precise and consistent assessments of protein expression patterns.

Despite these advances, Western blotting and IHC have some challenges when using them. One major difficulty is the possibility of nonspecific binding or cross-reactivity of antibodies, which can result in false-positive results or misunderstanding of data. This problem emphasizes the need for extensive antibody validation and the use of adequate controls in both procedures. Furthermore, factors such as tissue fixation, antigen retrieval, and detection system selection can all have an impact on the repeatability of IHC data, necessitating the use of defined protocols to assure consistency between experiments (Alberts et al., 2015; Han et al., 2008).

10.7. CRISPR-CAS9 GENE EDITING TECHNIQUES

Clustered regularly interspaced short palindromic repeat (CRISPR) and their associated protein (Cas-9) gene editing is a revolutionary advance in molecular biology, particularly in cell biochemistry. The CRISPR-Cas9 system is based on a naturally existing defense mechanism found in bacteria, which use this system to fend off viral infections by cutting the DNA of invading viruses. Scientists have taken advantage of this system's accuracy, transforming it into a powerful tool for targeting DNA sequences within genomes of organisms. This skill has far-reaching consequences for cell biochemistry since it allows researchers to modify genetic material with hitherto impossible accuracy and efficiency. One of CRISPR-Cas9's most important contributions to cell biochemistry is its role in gene function research. Understanding the function of certain genes is critical to biochemistry since genes encode proteins that carry out almost all biological functions. CRISPR-Cas9 allows for the development of gene knockouts, in which a certain gene is entirely deactivated. Scientists can deduce the significance of damaged genes in diverse cellular pathways by studying their phenotypic implications. This technique has proven particularly useful in dissecting complex biochemical processes like signal transduction, in which several genes frequently interact to modulate cellular responses (Gomperts et al., 2009).

In addition to knockout studies, CRISPR-Cas9 allows for more complex genetic modifications, including the insertion of particular mutations that resemble those present in human diseases. This technique is invaluable for developing cellular models of diseases, which are required to understand disease mechanisms at the molecular level. For example, by introducing mutations into genes important in metabolic pathways, researchers can examine the subsequent biochemical changes and find prospective therapeutic targets. This method has been used to represent disorders such as cystic fibrosis, in which mutations in the CFTR gene result in faulty chloride ion transport, and sickle cell anemia, which is caused by a single point mutation in the hemoglobin gene (Alberts et al., 2015).

CRISPR-Cas9 has also transformed the research of protein function in the setting of cell biology. Proteins are the cell's workhorses, catalyzing metabolic reactions, delivering signals, and keeping the cell structure intact. Researchers can investigate how changes in protein structure affect their function by introducing mutations into the genes that encode these proteins using CRISPR-Cas9. This is especially relevant for enzymes, which are proteins that catalyze biological reactions. Mutating certain amino acids within an enzyme can aid in the identification of residues required for catalytic activity. Furthermore, CRISPR-Cas9 can be used to analyze protein-protein interactions by deleting the genes that encode interaction partners, allowing researchers to more precisely map complicated biochemical networks (Baynes and Dominiczak, 2009).

Another key use of CRISPR-Cas9 in cell biochemistry is the study of epigenetic control. Epigenetics refers to changes in gene expression that do not entail modifications to the underlying DNA sequence. These alterations are frequently caused by chemical modifications to DNA or histone proteins, which can be passed along through cell division. CRISPR-Cas9 has been used for epigenetic research by combining the Cas9 protein with transcriptional activators or repressors. This enables researchers to control gene expression in a controlled manner, activating or silencing individual genes without changing their DNA sequence. This approach, known as CRISPRa (CRISPR activation) or CRISPRi (CRISPR interference), has given fresh insights into the regulatory networks that control gene expression in many biological situations. It also has the potential to develop novel therapeutic techniques for disorders involving dysregulated gene expression, such as cancer (Punekar, 2018).

One of the most intriguing features of CRISPR-Cas9 technology is its potential for therapeutic applications. Through the correction of mutations causing disease at the DNA level, CRISPR-Cas9 presents a promising treatment option for genetic disorders that were previously incurable. For instance, in animal models, researchers have used CRISPR-Cas9 to fix abnormalities in the dystrophin gene, which causes Duchenne muscular dystrophy. This accomplishment increases the likelihood of creating gene therapies that can treat or even eliminate hereditary illnesses in people. However, there are still a lot of obstacles to overcome before CRISPR-Cas9 may be used therapeutically, especially when it comes to specificity and transport. For these treatments to be safe and effective, it is essential that the CRISPR-Cas9 components enter the target cells and modify only the targeted DNA sequences (Davis, 2012).

The combination of CRISPR-Cas9 with other modern biochemistry techniques has increased its value. For example, combining CRISPR-Cas9 and single-cell RNA sequencing (scRNA-seq) enables researchers to delete genes while also profiling the effects of these modifications on gene expression in individual cells. This method offers a full perspective of the cellular response to genetic changes, making it very valuable for analyzing diverse cell populations, such as those present in malignancies. Researchers can uncover major drivers of cancer progression and prospective therapeutic targets by knocking off genes in cancer cells using CRISPR-Cas9 and then analyzing the ensuing gene expression profiles using scRNA-seq (Sharma et al., 2020).

Furthermore, the introduction of CRISPR-Cas9 in conjunction with sophisticated imaging techniques has created new opportunities for researching DNA repair processes. Live-cell imaging, for example, can be used to track the real-time dynamics of CRISPR-Cas9-mediated DNA double-strand break (DSB) repair. This has revealed new details about the methods by which cells detect and repair DNA damage, a procedure critical for genome stability. Understanding these repair processes is critical not just for basic research, but also for increasing the safety of CRISPR-based therapeutics, as unintentional off-target effects might result in deleterious mutations (Fu et al., 2013).

CRISPR-Cas9 technique in cell biology also has several limits which pose a great challenge when using it. One of the primary concerns is the possibility of off-target effects, in which the Cas9 protein causes double-strand breaks at undesired regions in the genome. These off-target effects might result in undesired alterations, which can skew experimental results or have negative consequences in therapeutic applications. To overcome this issue, researchers created high-fidelity Cas9 protein variants that had less off-target activity. These modified Cas9 variants preserve the capacity to target specific DNA sequences while reducing the chance of unintentional cuts. However, complete specificity remains a hurdle, and continuing research focuses on enhancing CRISPR-Cas9 precision (Fu et al., 2013).

Another disadvantage of CRISPR-Cas9 is that it relies on the cell's own DNA repair processes, specifically non-homologous end joining (NHEJ) and homology-directed repair (HDR). While NHEJ is the most common repair process, it frequently produces tiny insertions or deletions that might damage the gene at the target spot. In contrast, HDR can be used to induce precise genetic modifications using a donor DNA template, but it is less efficient and only happens during certain stages of the cell cycle. Efforts to improve HDR efficiency, such as the discovery of small compounds that promote HDR or the use of CRISPR-Cas9 variations that prefer HDR to NHEJ, are underway. These advances have the potential to broaden the variety of CRISPR-Cas9 applications in gene editing (Davis, 2012).

CRISPR-Cas9 has had a significant impact on the field of cell biochemistry by providing a diverse and precise tool for genome editing. Its uses span from basic research, which has improved our understanding of gene and protein function, to the creation of innovative therapeutic approaches for genetic illnesses. The combination of CRISPR-Cas9 with other biochemical tools, such as single-cell RNA sequencing and improved imaging, has increased its usefulness, allowing researchers to investigate cellular processes in unprecedented depth. As technology advances, addressing issues such as off-target effects and enhancing the efficiency of DNA repair mechanisms will be critical to realizing CRISPR-Cas9's full promise in both scientific and therapeutic settings.

Bibliography

1. Abubakar, A., & Haque, M. (2020). Preparation of medicinal plants: Basic extraction and fractionation procedures for experimental purposes. *Journal of Pharmacy and Bioallied Sciences, 12*(1), 1. https://doi.org/10.4103/jpbs.JPBS_175_19.
2. Akram, M. (2014). Citric acid cycle and role of its intermediates in metabolism. *Cell Biochemistry and Biophysics, 68*(3), 475–478. https://doi.org/10.1007/s12013-013-9750-1.
3. Alberts, B., Bray, D., Hopkin, K., Johnson, A. D., Lewis, J., Raff, M., Roberts, K., & Walter, P. (2015). *Essential Cell Biology*. Garland Science.
4. Angumeenal, A. R., & Venkappayya, D. (2013). An overview of citric acid production. *LWT–Food Science and Technology, 50*(2), 367–370. https://doi.org/10.1016/j.lwt.2012.05.016.
5. Baker, D. H. (2009). Advances in protein–amino acid nutrition of poultry. *Amino Acids, 37*(1), 29–41. https://doi.org/10.1007/s00726-008-0198-3.
6. Barbu, M. G., Condrat, C. E., Thompson, D. C., Bugnar, O. L., Cretoiu, D., Toader, O. D., Suciu, N., & Voinea, S. C. (2020). MicroRNA involvement in signaling pathways during viral infection. *Frontiers in Cell and Developmental Biology, 8*, 143. https://doi.org/10.3389/fcell.2020.00143.
7. Barnum, K. J., & O'Connell, M. J. (2014). Cell cycle regulation by checkpoints. In E. Noguchi & M. C. Gadaleta (Eds.), *Cell Cycle Control* (Vol. *1170*, pp. 29–40). Springer New York. https://doi.org/10.1007/978-1-4939-0888-2_2.
8. Baynes, J., & Dominiczak, M. H. (2009). *Medical Biochemistry*. Elsevier Health Sciences.
9. BBC Bitesize. (2024). What are plant and animal cells?–BBC Bitesize. Retrieved from: https://www.bbc.co.uk/bitesize/articles/zkm7wnb

(accessed on 8 November 2024).

10. Becker, P. B., & Workman, J. L. (2013). Nucleosome remodeling and epigenetics. *Cold Spring Harbor Perspectives in Biology, 5*(9), a017905. https://doi.org/10.1101/cshperspect.a017905.
11. Bhattacharyya, R. P., Reményi, A., Yeh, B. J., & Lim, W. A. (2006). Domains, motifs, and scaffolds: The role of modular interactions in the evolution and wiring of cell signaling circuits. *Annual Review of Biochemistry, 75*, 655–680. https://doi.org/10.1146/annurev.biochem.75.103004.142710.
12. Blankenberg, S., Rupprecht, H. J., Bickel, C., Peetz, D., Hafner, G., Tiret, L., Meyer, J., & for the AtheroGene Investigators. (2001). Circulating cell adhesion molecules and death in patients with coronary artery disease. *Circulation, 104*(12), 1336–1342. https://doi.org/10.1161/hc3701.095949.
13. Blumwald, E., Aharon, G. S., & Apse, M. P. (2000). Sodium transport in plant cells. *Biochimica et Biophysica Acta (BBA)–Biomembranes, 1465*(1), 140–151. https://doi.org/10.1016/S0005-2736(00)00135-8.
14. Boal, D. H. (2012). *Mechanics of the Cell*. Cambridge University Press.
15. Bonner, J., & Varner, J. E. (2012). *Plant Biochemistry*. Elsevier.
16. Boutros, T., Chevet, E., & Metrakos, P. (2008). Mitogen-activated protein (MAP) kinase/MAP kinase phosphatase regulation: Roles in cell growth, death, and cancer. *Pharmacological Reviews, 60*(3), 261–310. https://doi.org/10.1124/pr.107.00106.
17. Branzei, D., & Foiani, M. (2008). Regulation of DNA repair throughout the cell cycle. *Nature Reviews Molecular Cell Biology, 9*(4), 297–308. https://doi.org/10.1038/nrm2351.
18. Breckel, C. A., & Hochstrasser, M. (2021). Ubiquitin ligase redundancy and nuclear-cytoplasmic localization in yeast protein quality control. *Biomolecules, 11*(12), 1821. https://doi.org/10.3390/biom11121821.
19. Briscoe, J., & Thérond, P. P. (2013). The mechanisms of Hedgehog signalling and its roles in development and disease. *Nature Reviews Molecular Cell Biology, 14*(7), 416–429. https://doi.org/10.1038/nrm3598.
20. Brown, A. (2020). 4.2: A closer look at carbohydrates. *LibreTexts Nutrition*. Retrieved from: https://courses.lumenlearning.com/suny-mcc-ltnutrition/chapter/4-2-a-closer-look-at-carbohydrates/ (accessed on 8 November 2024).
21. BYJUS. (2023). What is DNA?–Meaning, DNA types, structure and functions. Retrieved from: https://byjus.com/biology/dna-structure/ (accessed on 8 November 2024).
22. BYJUS. (2024). Cell organelles–Types, structure and their functions.

Retrieved from: https://byjus.com/biology/cell-organelles/ (accessed on 8 November 2024).

23. Carreras, H. Z. (2024). An introduction to protein purification: Methods, technologies and applications. *Analysis & Separations from Technology Networks*. https://www.technologynetworks.com/analysis/articles/an-introduction-to-protein-purification-methods-technologies-and-applications-388443 (accessed on 8 November 2024).
24. Chagin, V. O., Stear, J. H., & Cardoso, M. C. (2010). Organization of DNA replication. *Cold Spring Harbor Perspectives in Biology, 2*(4), a000737. https://doi.org/10.1101/cshperspect.a000737.
25. Chandel, N. S. (2021). Glycolysis. *Cold Spring Harbor Perspectives in Biology, 13*(5), a040535. https://doi.org/10.1101/cshperspect.a040535.
26. Chhichholiya, Y., Singh, H. V., Vashistha, R., Singh, S., & Munshi, A. (2024). Deciphering the role of KRAS gene in oncogenesis: Focus on signaling pathways, genetic alterations in 3'UTR, KRAS specific miRNAs and therapeutic interventions. *Critical Reviews in Oncology/Hematology, 194*, 104250. https://doi.org/10.1016/j.critrevonc.2023.104250.
27. Chourpiliadis, C., & Mohiuddin, S. S. (2024). Biochemistry, gluconeogenesis. In *StatPearls*. StatPearls Publishing. http://www.ncbi.nlm.nih.gov/books/NBK544346/ (accessed on 8 November 2024).
28. Csordás, G., Renken, C., Várnai, P., Walter, L., Weaver, D., Buttle, K. F., Balla, T., Mannella, C. A., & Hajnóczky, G. (2006). Structural and functional features and significance of the physical linkage between ER and mitochondria. *The Journal of Cell Biology, 174*(7), 915–921. https://doi.org/10.1083/jcb.200604016.
29. Davis, L. (2012). *Basic Methods in Molecular Biology*. Elsevier.
30. Deegan, T. D., & Diffley, J. F. (2016). MCM: One ring to rule them all. *Current Opinion in Structural Biology, 37*, 145–151. https://doi.org/10.1016/j.sbi.2016.01.014.
31. Deretic, V., Saitoh, T., & Akira, S. (2013). Autophagy in infection, inflammation, and immunity. *Nature Reviews Immunology, 13*(10), 722–737. https://doi.org/10.1038/nri3532.
32. Diamante, L., & Martello, G. (2022). Metabolic regulation in pluripotent stem cells. *Current Opinion in Genetics & Development, 75*, 101923. https://doi.org/10.1016/j.gde.2022.101923.
33. Du, X., Li, Y., Xia, Y. L., Ai, S. M., Liang, J., Sang, P., Ji, X. L., & Liu, S. Q. (2016). Insights into protein–ligand interactions: Mechanisms, models, and methods. *International Journal of Molecular Sciences, 17*(2), 144. https://doi.org/10.3390/ijms17020144.

34. Ezell, O. (2024). Eukaryotic RNA processing & splicing–MCAT biology. MedSchoolCoach. MedSchoolCoach. Helping You Achieve Your Medical School Dreams. Retrieved from: https://www.medschoolcoach.com/eukaryotic-rna-processing-splicing-mcat-biology/ (accessed on 8 November 2024).
35. Feingold, K. R. (2024). Introduction to lipids and lipoproteins. In *Endotext* [Internet]. MDText.com, Inc. https://www.ncbi.nlm.nih.gov/sites/books/NBK305896/ (accessed on 8 November 2024).
36. Finn, R. S., Martin, M., Rugo, H. S., Jones, S., Im, S. A., Gelmon, K., Harbeck, N., Lipatov, O. N., Walshe, J. M., Moulder, S., Gauthier, E., Lu, D. R., Randolph, S., Diéras, V., & Slamon, D. J. (2016). Palbociclib and letrozole in advanced breast cancer. *New England Journal of Medicine, 375*(20), 1925–1936. https://doi.org/10.1056/NEJMoa1607303.
37. Forrest, L. R., Krämer, R., & Ziegler, C. (2011). The structural basis of secondary active transport mechanisms. *Biochimica et Biophysica Acta (BBA)–Bioenergetics, 1807*(2), 167–188. https://doi.org/10.1016/j.bbabio.2010.10.014.
38. Furgason, J. M., & Bahassi, E. M. (2013). Targeting DNA repair mechanisms in cancer. *Pharmacology & Therapeutics, 137*(3), 298–308. https://doi.org/10.1016/j.pharmthera.2012.10.009.
39. Gadsby, D. C. (2009). Ion channels versus ion pumps: The principal difference, in principle. *Nature Reviews Molecular Cell Biology, 10*(5), 344–352. https://doi.org/10.1038/nrm2668.
40. Gagneux, P., Hennet, T., & Varki, A. (2022). Biological functions of glycans. In A. Varki, R. D. Cummings, J. D. Esko, P. Stanley, G. W. Hart, M. Aebi, D. Mohnen, T. Kinoshita, N. H. Packer, J. H. Prestegard, R. L. Schnaar, & P. H. Seeberger (Eds.), *Essentials of Glycobiology* (4th ed.). Cold Spring Harbor Laboratory Press. http://www.ncbi.nlm.nih.gov/books/NBK579984/ (accessed on 8 November 2024).
41. García-Segura, L., Pérez-Andrade, M., & Miranda-Ríos, J. (2013). The emerging role of microRNAs in the regulation of gene expression by nutrients. *Lifestyle Genomics, 6*(1), 16–31. https://doi.org/10.1159/000345826.
42. GeeksforGeeks. (2024). Cell signaling. *GeeksforGeeks*. https://www.geeksforgeeks.org/cell-signaling/ (accessed on 8 November 2024).
43. George, A. A., Schiltz, R. L., & Hager, G. L. (2009). Dynamic access of the glucocorticoid receptor to response elements in chromatin. *The International Journal of Biochemistry & Cell Biology, 41*(1), 214–224. https://doi.org/10.1016/j.biocel.2008.09.019.

44. Gomperts, B. D., Kramer, I. M., & Tatham, P. E. R. (2009). *Signal Transduction*. Academic Press.
45. Guengerich, F. P. (2023). Mechanisms of enzyme catalysis and inhibition. In *Reference Module in Biomedical Sciences*. Elsevier. https://doi.org/10.1016/B978-0-323-95488-4.00008-5.
46. Gupta, V. K., & Chaudhuri, O. (2022). Mechanical regulation of cell-cycle progression and division. *Trends in Cell Biology, 32*(9), 773–785. https://doi.org/10.1016/j.tcb.2022.03.010.
47. Ha, M. (2013). Understanding the chromatin remodeling code. *Plant Science, 211*, 137–145. https://doi.org/10.1016/j.plantsci.2013.07.006.
48. Han, X., Aslanian, A., & Yates, J. R. (2008). Mass spectrometry for proteomics. *Current Opinion in Chemical Biology, 12*(5), 483–490. https://doi.org/10.1016/j.cbpa.2008.07.024.
49. Hartl, F. U., Bracher, A., & Hayer-Hartl, M. (2011). Molecular chaperones in protein folding and proteostasis. *Nature, 475*(7356), 324–332. https://doi.org/10.1038/nature10317.
50. Hayashi, M. T., & Karlseder, J. (2013). DNA damage associated with mitosis and cytokinesis failure. *Oncogene, 32*(39), 4593–4601. https://doi.org/10.1038/onc.2012.615.
51. Heim, M., Römer, L., & Scheibel, T. (2010). Hierarchical structures made of proteins: The complex architecture of spider webs and their constituent silk proteins. *Chemical Society Reviews, 39*(1), 156–164. https://doi.org/10.1039/b813273a.
52. Helena, J., Joubert, A., Grobbelaar, S., Nolte, E., Nel, M., Pepper, M., Coetzee, M., & Mercier, A. (2018). Deoxyribonucleic acid damage and repair: Capitalizing on our understanding of the mechanisms of maintaining genomic integrity for therapeutic purposes. *International Journal of Molecular Sciences, 19*(4), 1148. https://doi.org/10.3390/ijms19041148.
53. Hohmann-Marriott, M. F., & Blankenship, R. E. (2011). Evolution of photosynthesis. *Annual Review of Plant Biology, 62*, 515–548. https://doi.org/10.1146/annurev-arplant-042110-103811.
54. Hunter, N. (2015). Meiotic recombination: The essence of heredity. *Cold Spring Harbor Perspectives in Biology*, a016618. https://doi.org/10.1101/cshperspect.a016618.
55. Inzé, D., & De Veylder, L. (2006). Cell cycle regulation in plant development. *Annual Review of Genetics, 40*(1), 77–105. https://doi.org/10.1146/annurev.genet.40.110405.090431.

56. Jain, S., Venkataraman, A., Wechsler, M. E., & Peppas, N. A. (2021). Messenger RNA-based vaccines: Past, present, and future directions in the context of the COVID-19 pandemic. *Advanced Drug Delivery Reviews, 179*, 114000. https://doi.org/10.1016/j.addr.2021.114000.

57. Jang, K. H., Heras, C. R., & Lee, G. (2022). m6A in the signal transduction network. *Molecules and Cells, 45*(7), 435–443. https://doi.org/10.14348/molcells.2022.0017.

58. Janson, J. C. (2011). *Protein Purification: Principles, High Resolution Methods, and Applications*. John Wiley & Sons.

59. Jin, S., Borkhuu, O., Bao, W., & Yang, Y. T. (2016). Signaling pathways in thyroid cancer and their therapeutic implications. *Journal of Clinical Medicine Research, 8*(4), 284–296. https://doi.org/10.14740/jocmr2480w.

60. Joo, C., Balci, H., Ishitsuka, Y., Buranachai, C., & Ha, T. (2008). Advances in single-molecule fluorescence methods for molecular biology. *Annual Review of Biochemistry, 77*(1), 51–76. https://doi.org/10.1146/annurev.biochem.77.070606.101543.

61. Kabani, M., Pilard, M., & Melki, R. (2020). Glucose availability dictates the export of the soluble and prion forms of Sup35p via periplasmic or extracellular vesicles. *Molecular Microbiology, 114*(2), 322–332. https://doi.org/10.1111/mmi.14515.

62. Keller, M. A., Piedrafita, G., & Ralser, M. (2015). The widespread role of non-enzymatic reactions in cellular metabolism. *Current Opinion in Biotechnology, 34*, 153–161. https://doi.org/10.1016/j.copbio.2014.12.020.

63. Kim, J., Kundu, M., Viollet, B., & Guan, K. L. (2011). AMPK and mTOR regulate autophagy through direct phosphorylation of Ulk1. *Nature Cell Biology, 13*(2), 132–141. https://doi.org/10.1038/ncb2152.

64. Kornblihtt, A. R., Schor, I. E., Alló, M., Dujardin, G., Petrillo, E., & Muñoz, M. J. (2013). Alternative splicing: A pivotal step between eukaryotic transcription and translation. *Nature Reviews Molecular Cell Biology, 14*(3), 153–165. https://doi.org/10.1038/nrm3525.

65. Kornbrust, B. A., Forman, T., & Matveeva, I. (2012). Applications of enzymes in breadmaking. In S. P. Cauvain (Ed.), *Breadmaking* (2nd ed., pp. 470–498). Woodhead Publishing. https://doi.org/10.1533/9780857095695.2.470.

66. Koutroumpas, K., & Lygeros, J. (2011). Modeling and analysis of DNA replication. *Automatica, 47*(6), 1156–1164. https://doi.org/10.1016/j.automatica.2011.02.007.

67. Kouzarides, T. (2007). Chromatin modifications and their function. *Cell, 128*(4), 693–705. https://doi.org/10.1016/j.cell.2007.02.005.

68. Lenk, C. (2017). *Role of Coupling Conditions for Pattern Formation in Excitable Media*. http://books.google.ie/books?id=2G940AEACAAJ (accessed on 8 November 2024).

69. Leonard, A. C., & Mechali, M. (2013). DNA replication origins. *Cold Spring Harbor Perspectives in Biology, 5*(10), a010116. https://doi.org/10.1101/cshperspect.a010116.

70. Libretexts. (2022). The two parts of photosynthesis. In *Biology LibreTexts*. https://bio.libretexts.org/Bookshelves/Microbiology/Microbiology_%28Boundless%29/05%3A_Microbial_Metabolism/5.11%3A_Phototrophy/5.11C%3A_The_Two_Parts_of_Photosynthesis (accessed on 8 November 2024).

71. Licatalosi, D. D., & Darnell, R. B. (2010). RNA processing and its regulation: Global insights into biological networks. *Nature Reviews Genetics, 11*(1), 75–87. https://doi.org/10.1038/nrg2673.

72. Litan, A., & Langhans, S. A. (2015). Cancer as a channelopathy: Ion channels and pumps in tumor development and progression. *Frontiers in Cellular Neuroscience, 9*, 86. https://doi.org/10.3389/fncel.2015.00086.

73. Lodewijk, I., Nunes, S. P., Henrique, R., Jerónimo, C., Dueñas, M., & Paramio, J. M. (2021). Tackling tumor microenvironment through epigenetic tools to improve cancer immunotherapy. *Clinical Epigenetics, 13*(1), 46. https://doi.org/10.1186/s13148-021-01046-0.

74. Lopez-Barbosa, N., Ludwicki, M. B., & DeLisa, M. P. (2019). Proteome editing using engineered proteins that hijack cellular quality control machinery. *AIChE Journal, 66*(3). https://doi.org/10.1002/aic.16854.

75. Mackiewicz, M., Shockley, K. R., Romer, M. A., Galante, R. J., Zimmerman, J. E., Naidoo, N., Baldwin, D. A., Jensen, S. T., Churchill, G. A., & Pack, A. I. (2007). Macromolecule biosynthesis: A key function of sleep. *Physiological Genomics, 31*(3), 441–457. https://doi.org/10.1152/physiolgenomics.00275.2006.

76. Mahoney, D. E., Hiebert, J. B., Thimmesch, A., Pierce, J. T., Vacek, J. L., Clancy, R. L., Sauer, A. J., & Pierce, J. D. (2018). Understanding D-Ribose and mitochondrial function. *Advances in Bioscience and Clinical Medicine, 6*(1), 1. https://doi.org/10.7575/aiac.abcmed.v.6n.1p.1.

77. Marchetti, R., Perez, S., Arda, A., Imberty, A., Jimenez-Barbero, J., Silipo, A., & Molinaro, A. (2016). "Rules of engagement" of protein-glycoconjugate interactions: A molecular view achievable by using NMR spectroscopy and molecular modeling. *ChemistryOpen, 5*(4), 274–296. https://doi.org/10.1002/open.201600024.

78. Mardamshina, M., & Geiger, T. (2017). Next-generation proteomics and its application to clinical breast cancer research. *American Journal of Pathology, 187*(10), 2175–2184. https://doi.org/10.1016/j.ajpath.2017.07.003.

79. Marshall, B. T., Long, M., Piper, J. W., Yago, T., McEver, R. P., & Zhu, C. (2003). Direct observation of catch bonds involving cell-adhesion molecules. *Nature, 423*(6936), 190–193. https://doi.org/10.1038/nature01605.

80. Matsui, W. H. (2016). Cancer stem cell signaling pathways. *Medicine, 95*(1S), S8–S19. https://doi.org/10.1097/MD.0000000000004765.

81. Mattaini, K. (2020a). Chapter 8. Membrane transport. In *Pressbooks*. https://rwu.pressbooks.pub/bio103/chapter/membrane-transport/ (accessed on 8 November 2024).

82. Mattaini, K. (2020b). Chapter 9. Cell Communication. *Pressbooks*. https://rwu.pressbooks.pub/bio103/chapter/cell-communication/ (accessed on 8 November 2024).

83. Melkonian, E. A., Asuka, E., & Schury, M. P. (2024). Physiology, gluconeogenesis. In *StatPearls*. StatPearls Publishing. http://www.ncbi.nlm.nih.gov/books/NBK541119/ (accessed on 8 November 2024).

84. Mollicone, R., Moore, S. E. H., Bovin, N., Garcia-Rosasco, M., Candelier, J. J., Martinez-Duncker, I., & Oriol, R. (2009). Activity, splice variants, conserved peptide motifs, and phylogeny of two new α1,3-fucosyltransferase families (FUT10 and FUT11). *Journal of Biological Chemistry, 284*(7), 4723–4738. https://doi.org/10.1074/jbc.M809312200.

85. Mukherjee, S. (2023). Cell cycle: Definition, phases, and diagram. *Science Facts*. https://www.sciencefacts.net/cell-cycle.html (accessed on 8 November 2024).

86. Nair, K. S., Naidoo, R., & Chetty, R. (2005). Expression of cell adhesion molecules in oesophageal carcinoma and its prognostic value. *Journal of Clinical Pathology, 58*(4), 343–351. https://doi.org/10.1136/jcp.2004.018036.

87. Newton, A. C., Bootman, M. D., & Scott, J. D. (2016). Second messengers. *Cold Spring Harbor Perspectives in Biology, 8*(8), a005926. https://doi.org/10.1101/cshperspect.a005926.

88. Nucleic Acids Study Guide–Inspirit Learning Inc. (2023). *Inspirit Learning Inc–Experience the Future of Education*. https://inspiritvr.com/nucleic-acids-study-guide-2/ (accessed on 8 November 2024).

89. Nusse, R., & Clevers, H. (2017). Wnt/β-catenin signaling, disease, and

emerging therapeutic modalities. *Cell, 169*(6), 985–999. https://doi.org/10.1016/j.cell.2017.05.016.

90. O'Donnell, L. (2014). Mechanisms of spermiogenesis and spermiation and how they are disturbed. *Spermatogenesis, 4*(2), e979623. https://doi.org/10.4161/21565562.2014.979623.
91. Oh, N., & Park, J. H. (2014). Endocytosis and exocytosis of nanoparticles in mammalian cells. *International Journal of Nanomedicine, 9*(sup1), 51–63. https://doi.org/10.2147/IJN.S26592.
92. Pessarakli, M. (2024). *Handbook of Photosynthesis*. CRC Press.
93. Pfeiffer, V., & Lingner, J. (2013). Replication of telomeres and the regulation of telomerase. *Cold Spring Harbor Perspectives in Biology, 5*(5), a010405. https://doi.org/10.1101/cshperspect.a010405.
94. Phillips, M. J., & Voeltz, G. K. (2016). Structure and function of ER membrane contact sites with other organelles. *Nature Reviews Molecular Cell Biology, 17*(2), 69–82. https://doi.org/10.1038/nrm.2015.8.
95. Pigman, W. (2012). *The Carbohydrates: Chemistry, Biochemistry, Physiology*. Elsevier.
96. Pollard, T. D., Earnshaw, W. C., Lippincott-Schwartz, J., & Johnson, G. (2022). *Cell Biology: Cell Biology*. Elsevier Health Sciences.
97. Puiu, T. (2023). What are the steps of DNA replication? *ZME Science*. https://www.zmescience.com/medicine/genetic/dna-replication-steps-43264/ (accessed on 8 November 2024).
98. Punekar, N. S. (2018). *Enzymes: Catalysis, Kinetics and Mechanisms*. Springer.
99. Ramazi, S., & Zahiri, J. (2021). Posttranslational modifications in proteins: Resources, tools and prediction methods. *Database: The Journal of Biological Databases and Curation, 2021*, baab012. https://doi.org/10.1093/database/baab012.
100. Ramírez-Clavijo, S., & Montoya-Ortíz, G. (2013). Gene expression and regulation. In *Autoimmunity: From Bench to Bedside* [Internet]. El Rosario University Press. https://www.ncbi.nlm.nih.gov/books/NBK459456/ (accessed on 8 November 2024).
101. Ransom, M., Dennehey, B. K., & Tyler, J. K. (2010). Chaperoning histones during DNA replication and repair. *Cell, 140*(2), 183–195. https://doi.org/10.1016/j.cell.2010.01.004.
102. Rönnblom, L., & Leonard, D. (2019). Interferon pathway in SLE: One key to unlocking the mystery of the disease. *Lupus Science & Medicine, 6*(1), e000270. https://doi.org/10.1136/lupus-2018-000270.

103. Saenger, W. (2013). *Principles of Nucleic Acid Structure*. Springer Science & Business Media.
104. Salentin, S., Schreiber, S., Haupt, V. J., Adasme, M. F., & Schroeder, M. (2015). PLIP: Fully automated protein–ligand interaction profiler. *Nucleic Acids Research, 43*(W1), W443–W447. https://doi.org/10.1093/nar/gkv315.
105. Satyanarayana, A., & Kaldis, P. (2009). Mammalian cell-cycle regulation: Several Cdks, numerous cyclins and diverse compensatory mechanisms. *Oncogene, 28*(33), 2925–2939. https://doi.org/10.1038/onc.2009.170.
106. Schaeffer, R. D., Kinch, L. N., Liao, Y., & Grishin, N. V. (2016). Classification of proteins with shared motifs and internal repeats in the ECOD database. *Protein Science: A Publication of the Protein Society, 25*(7), 1188–1203. https://doi.org/10.1002/pro.2893.
107. Sclafani, R. A., & Holzen, T. M. (2007). Cell cycle regulation of DNA replication. *Annual Review of Genetics, 41*, 237–280. https://doi.org/10.1146/annurev.genet.41.110306.130308.
108. Shay, J. W., & Wright, W. E. (2019). Telomeres and telomerase: Three decades of progress. *Nature Reviews Genetics, 20*(5), 299–309. https://doi.org/10.1038/s41576-019-0099-1.
109. Shuken, S. R. (2023). An introduction to mass spectrometry-based proteomics. *Journal of Proteome Research, 22*(7), 2151–2171. https://doi.org/10.1021/acs.jproteome.2c00838.
110. Söti, C., & Csermely, P. (2002). Chaperones and aging: Role in neurodegeneration and in other civilizational diseases. *Neurochemistry International, 41*(6), 383–389. https://doi.org/10.1016/S0197-0186(02)00043-8.
111. Srinivasan, B. (2020). Explicit treatment of non-Michaelis-Menten and atypical kinetics in early drug discovery. https://doi.org/10.20944/preprints202010.0179.v1.
112. Stein, W. (2012). *Transport and Diffusion Across Cell Membranes*. Elsevier.
113. Stollar, E. J., & Smith, D. P. (2020). Uncovering protein structure. *Essays in Biochemistry, 64*(4), 649–680. https://doi.org/10.1042/ebc20190042.
114. Tamang, S. (2023). Gene expression: Stages, regulations, methods. *Microbe Notes*. https://microbenotes.com/gene-expression/ (accessed on 8 November 2024).
115. van Meer, G., Voelker, D. R., & Feigenson, G. W. (2008). Membrane lipids: Where they are and how they behave. *Nature Reviews Molecular*

Cell Biology, 9(2), 112–124. https://doi.org/10.1038/nrm2330.

116. Van Roy, F., & Berx, G. (2008). The cell-cell adhesion molecule E-cadherin. *Cellular and Molecular Life Sciences, 65*(23), 3756–3788. https://doi.org/10.1007/s00018-008-8281-1.
117. Vignali, D. A. A., Collison, L. W., & Workman, C. J. (2008). How regulatory T cells work. *Nature Reviews Immunology, 8*(7), 523–532. https://doi.org/10.1038/nri2343.
118. Wenzel, E. S., & Singh, A. T. K. (2018). Cell-cycle checkpoints and aneuploidy on the path to cancer. *In Vivo, 32*(1), 1–5.
119. Whitford, D. (2013). *Proteins: Structure and Function*. John Wiley & Sons.
120. Wilkin, D., & Brainard, J. (2022). CK-12 Foundation. CK-12 Foundation. Retrieved from: https://flexbooks.ck12.org/cbook/ck-12-biology-flexbook-2.0/section/2.16/primary/lesson/exocytosis-and-endocytosis-bio/.
121. Wu, L. G., Hamid, E., Shin, W., & Chiang, H. C. (2014). Exocytosis and endocytosis: Modes, functions, and coupling mechanisms. *Annual Review of Physiology, 76*(1), 301–331. https://doi.org/10.1146/annurev-physiol-021113-170305.
122. Wyatt Shields IV, C., Reyes, C. D., & López, G. P. (2015). Microfluidic cell sorting: A review of the advances in the separation of cells from debulking to rare cell isolation. *Lab on a Chip, 15*(5), 1230–1249. https://doi.org/10.1039/C4LC01246A.

INDEX

G

H

I

K

L

M

T

V

W

X

Z

MIX
Papier | Fördert
gute Waldnutzung
FSC® C083411

Kolibri 360 GmbH
Ferdinand-Jühlke-Straße 7
99095 Erfurt, Deutschland
produktsicherheit@kolibri360.de